*Other books on servicing by Gordon J. King*

Colour Television Servicing
F.M. Radio Servicing Handbook
Hi-Fi and Tape Recorder Handbook
Newnes Colour Television Servicing Manual
Practical Aerial Handbook
Radio and Audio Servicing Handbook
Radio, Television and Audio Test Instruments
Rapid Servicing of Transistor Equipment
Television Servicing Handbook
The Audio Handbook

# Servicing with the
# OSCILLOSCOPE

# Servicing with the
# OSCILLOSCOPE

A GUIDE TO TESTING, ADJUSTMENT AND
FAULT LOCATION IN RADIO, TELEVISION AND AUDIO EQUIPMENT

## GORDON J. KING
T.ENG. (C.E.I.), R.TECH. ENG., A.M.I.E.R.E., F.I.S.T.C.,
F.S.C.T.E., M.I.P.R.E., M.A.E.S.

LONDON
NEWNES-BUTTERWORTHS

THE BUTTERWORTH GROUP

ENGLAND     Butterworth & Co (Publishers) Ltd
            London: 88 Kingsway, WC2B 6AB

AUSTRALIA   Butterworths Pty Ltd
            Sydney: 586 Pacific Highway, NSW 2067
            Melbourne: 343 Little Collins Street, 3000
            Brisbane: 240 Queen Street, 4000

CANADA      Butterworth & Co (Canada) Ltd
            Scarborough: 2265 Midland Avenue, Ontario M1P 4S1

NEW ZEALAND Butterworths of New Zealand Ltd
            Wellington: 26-28 Waring Taylor Street, 1

SOUTH AFRICA Butterworth & Co (South Africa) (Pty) Ltd
            Durban: 152-154 Gale Street

First published in 1969
Second edition 1976

© Gordon J. King, 1976

All rights reserved. No part of this publication may be reproduced or transmitted in any form or by any means, including photocopying and recording, without the written permission of the copyright holder, application for which should be addressed to the publisher. Such written permission must also be obtained before any part of this publication is stored in a retrieval system of any nature.

This book is sold subject to the Standard Conditions of Sale of Net Books and may not be sold in the UK below the net price given by the publishers in their current price list.

ISBN 0 408 00195 X

Printed Offset Litho and bound in England by
Cox & Wyman Ltd, London, Fakenham and Reading

# PREFACE

THE oscilloscope is undoubtedly a measuring and diagnostic instrument of prodigious versatility. Not only is it capable of performing with relative ease many of the jobs normally undertaken by analogue instruments, but simultaneously it reveals the precise nature of the signal applied to it. It will yield most information on inputs from d.c. to h.f., limited only by its amplifiers and timebase. It compares and measures the phase, time and frequency of recurring signals and traces out the response characteristics of tuned circuits and amplifiers with the help of a partnering wobbulator. Indeed, there are so many jobs it can do that it would require almost a book alone to describe them.

It has been my aim in this book to present the instrument as a valuable aid to servicing and fault-finding in radio, television and audio equipment, including the latest stereo radio and colour television circuits. I have endeavoured to write about its application essentially from the practical standpoint. The illustrations include many off-the-screen photographs showing the oscilloscope traces to expect in normal and faulty equipment, while in the television chapters these oscillograms are supplemented by off-the-screen television fault photographs. All these pictures were taken in my laboratory over several years of design, development and experimentation, and since they are real displays they bring the reader into direct contact with the c.r.t. screen.

With the development of colour television, the oscilloscope is assuming a greater importance in television servicing than hitherto. This is because colour sets carry a greater number of more complex signals than their monochrome counterparts and radio receivers, and also because many colour sets adopt the module technique, making it necessary to X-ray (so to speak) the various sections to discover the one responsible for a particular fault while the modules are *in situ*. Only the oscilloscope can X-ray the circuits and show abnormal signals in this way.

The oscilloscope, used with an audio sine-wave and square-wave generator, is essential for quality checks and fault-finding in hi-fi amplifiers, tape-recorders and other kinds of audio equipment, and a chapter is devoted to this subject.

PREFACE

In bringing this book up to date I have not only included more information on the oscilloscope as an instrument, detailing some of the more recent features, but also extended some of the chapters dealing with application, particularly with regard to measurements and tests in audio equipment and colour television. Some of the original off-screen oscillograms have been replaced by new ones, and additional displays have been included in most of the chapters to expose some of the less well known secrets of oscilloscope testing.

More emphasis has also been given to the test and measurement aspects, and, although the term 'servicing' has been retained in the title, the book will now also be found to have value to the technician whose job it is to evaluate the various items of electronic equipment. In any case, after a servicing operation it is generally required to check that the parameters of the manufacturer's specification have not been degraded, and there is no doubt that the oscilloscope is a deep-seeking tool in this respect as well as many others. Without it modern servicing and testing would be almost an impossibility.

Most of the oscillograms here presented were photographed directly from the screens of oscilloscopes employed in my own test laboratory with a Polaroid oscilloscope camera or 35 mm film, and photography information is given in the text where appropriate.

My thanks are again due to my wife Babs for her devotion in forming and typing the manuscript from my rough notes and for preparing the index. I would also like to thank Gordon J. King (Enterprises) Ltd for allowing the use of their test equipment and the various manufacturers without whose assistance the book could never have been written.

Brixham, Devon *Gordon J. King*

# CONTENTS

| | | |
|---|---|---|
| 1 | INTRODUCTION TO THE OSCILLOSCOPE | 1 |
| 2 | APPLYING THE OSCILLOSCOPE | 23 |
| 3 | VIDEO WAVEFORMS | 33 |
| 4 | SYNCHRONISING WAVEFORMS | 48 |
| 5 | TIMEBASE WAVEFORMS | 67 |
| 6 | TELEVISION TESTS FOR HUM, DISTORTION AND RESPONSE | 102 |
| 7 | VISUAL CIRCUIT ALIGNMENT | 115 |
| 8 | COLOUR TELEVISION WAVEFORMS | 132 |
| 9 | STEREO RADIO WAVEFORMS | 158 |
| 10 | TESTING AUDIO EQUIPMENT | 169 |
| | INDEX | 205 |

# 1: INTRODUCTION TO THE OSCILLOSCOPE

SIGNALS in television, radio and audio circuits – that is, currents and voltages which are changing and repetitive, whether they be in the receiving or the timebase sections – have amplitude, frequency and characteristic 'shape'. The oscilloscope translates these electrical qualities into visual symbols: it takes a whole series of 'instantaneous' changes – a slice of time – and lays it before us as a diagram. In more technical terms, the oscilloscope has the ability to capture, display and analyse a time-domain waveform.

There are other instruments which can reveal the amplitude and the frequency of signals; but the oscilloscope (scope for short) is the only one that also shows the rate-of-change, in particular the variation in this which gives a signal its special character – its 'shape'. By showing what is going on electrically over a short period of time, the scope virtually X-rays a circuit – and so eliminates guesswork in diagnosis.

Oscilloscopes employ electrostatic deflection of the c.r.t. beam (not magnetic deflection as is used in television receivers) and this means that, essentially, they indicate change of potential (voltage). However, by making the current in a circuit develop a voltage across a resistor, a voltage can be obtained for application to the scope, and this will vary like the current itself. Sometimes a resistor already in the circuit can be used for the take-off; otherwise a resistor can be inserted for the purpose, taking care to use a value low enough not to disturb too much the normal working of the circuit.

Being voltage-operated, the scope's c.r.t. is itself a very high-impedance device. On direct input to the deflection plates, the impedance is measured in megohms, associated with a little shunt and/or series capacitance.

For high-frequency signal observation the shunt capacitance must be kept very low to avoid signal attenuation and waveform distortion. For low-frequency applications any series capacitance incorporated (for d.c. blocking) must not be too small; and, of course, on d.c. tests such a capacitor must be shorted out.

**Simple application**

In its simplest application the scope is just a high-resistance voltmeter. The spot is deflected vertically on the screen, usually against a superimposed transparent scale or graticule, in proportion to the strength of the electrostatic field produced between the deflection plates by the applied voltage. These plates for vertical deflection (themselves actually horizontal) are known as the Y plates.

Beam deflection is generally linear, the amount of deflection being proportional to the volts of differential potential across the plates. With direct voltage across the Y plates, the beam and hence the spot on the screen moves up or down (according to the polarity of the input) to an 'off centre' position and stays there. With alternating voltage across the Y plates, the spot moves up and down, 'drawing' a line between the peak voltage on one half-cycle and the peak voltage on the other half-cycle, thus spanning the peak-to-peak value, as shown in Fig. 1.1.

*Deflection* or *sensitivity factor* ($F_s$) is governed by various c.r.t. parameters, such as the distance between the tube end of the deflection plates and the screen ($d_1$), the overall length of the plates ($p_1$), the distance between the surfaces of the plates ($d_2$) and the potential of the electron beam ($V_k$). Since the deflection ($D$) is directly proportional to the deflection voltage differential (i.e., $V_1 - V_2$), the sensitivity factor can be expressed as

$$F_s = \frac{V_1 - V_2}{D} = \frac{K V_k d_2}{p_1 d_1}$$

where $K$ is a constant related to the postaccelerator field of the electron gun.

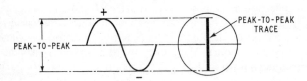

*Fig. 1.1.* With an alternating Y input the spot moves up and down, giving a vertical line display.

**Y amplifier**

The sensitivity factor of the c.r.t. is thus intrinsically low, so it is usually necessary to amplify the signal for display before it is applied to the Y plates. A Y amplifier with a gain control (i.e. amplifier attenuator) is incorporated in the instrument between the input terminals and Y plates. Sensitivity is expressed in 'volts per centimetre' (V/cm) or volts per division (V/div), the graticule in front of the screen being suitably ruled to allow the measurement of vertical deflection.

To cater for a reasonable range of signal voltages a Y amplifier attenuator control is necessary. This may be a rotary switch, a continuous control, or both. A Y range from about 300 V/cm (depending on the Y amplifier) down to at least 100 mV/cm over the maximum bandwidth of the Y amplifier is often a basic requirement. An even greater sensitivity is demanded for certain applications, including low-level audio work and colour television, such as 10 mV/cm or less. Some Y amplifiers are equipped with switching to increase the sensitivity over that established by the main attenuators by factors of 10 and 100. However, in the highest sensitivity setting the bandwidth of the Y amplifier is often curtailed, and this must be taken into account when fast-transient types of signal are being examined.

Deflection on the horizontal or X axis can be obtained in like manner by applying a voltage across the other pair of plates of the c.r.t.

**Timebase**

Display of television waveforms usually requires the X plates to be connected to a timebase (a sawtooth oscillator and amplifier) built into the instrument. This serves to deflect the spot linearly at a preset speed from the left- to the right-hand side of the screen, followed by rapid flyback for the start of the next trace. During the retrace many scopes black out (bias off) the spot and this is a desirable feature for some applications.

Timebase repetition frequencies above about 10 Hz turn the spot into the appearance of a line. This is because of the afterglow of the screen and the effect of persistence of vision. Similarly a vertical line is produced by a signal above about 10 Hz applied to the Y input (see Fig. 1.2).

When the timebase speed is such that the spot covers the screen width and flies back in the same time as the signal on the Y plates takes to go through one complete cycle, then the spot will trace the waveform of the Y-input cycle, and successive traces will overlay each other to give a steady display If the timebase is set to a lower speed, more cycles will be shown – but only if the horizontal scan always begins at the same point of the signal waveform which is fed to the Y input. That is, the signal frequency must be an exact multiple of the timebase frequency.

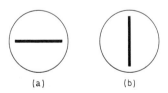

*Fig. 1.2.* At frequencies above about 10 Hz a horizontal line appears on the screen (*a*) and a vertical line (*b*) when the signal is applied to the X and Y inputs respectively.

# SERVICING WITH THE OSCILLOSCOPE

When the timebase frequency is anything less than the period of one signal cycle, or there is not a whole number of signal repetitions during a timebase scan, then the traces do not overlap each other and confusion results on the screen.

Timebases of all scopes used to be calibrated in repetition frequency but the trend now is to use 'sweep speed' calibration. This gives the time it takes for the spot to sweep a centimetre of screen. Such a small deflection takes only micro or milliseconds at ordinary timebase frequencies. The speed control is thus calibrated in μs/cm and ms/cm. A sweep range suitable for television applications is about 100 ms/cm to 1 μs/cm by means of about ten switch positions. In addition to the range switch some scopes have 'fine sweep' control, consisting of a variable potentiometer to swing the selected sweep speed a little one way or the other.

A 50 Hz mains supply cycle takes 20 ms. With the timebase adjusted to sweep the screen in 20 ms a complete 50 Hz sine wave will thus be displayed. If the screen diameter is, say, 6 cm, the sweep control needs to be set to about 3·3 ms/cm. In practice the extreme ends of the trace are cut off, due to the action of the scope synchronising, so it is as well always to run the sweep so that at least two cycles (one complete and the other almost complete) are displayed.

Whether or not a scope will display two or more waveforms depends on the frequency of the test signal and the maximum sweep speed of the scope. A scope with a maximum sweep of 1 μs/cm will trace one complete 1 MHz waveform over each cm of X axis. Six waveforms (that is, five complete and one almost complete) will appear on a 6 cm screen.

Time in seconds taken by a cycle of signal is the reciprocal of the frequency. A 2 Hz signal takes half a second (500 ms), a 10 Hz takes 100 ms and so on. Time in ms or μs is obtained (respectively) by dividing 1,000 or 1,000,000 by the signal frequency in Hz.

We tend to think of speed and frequency as synonymous but, in fact, amplitude is also involved. With a scope, the frequency of the X timebase can be kept constant but the amplitude of deflection increased – the spot having to travel faster – until a large part of the electron-beam movement has no effect and only a portion of the sweep appears on the screen.

**Horizontal gain**

This trace magnification, or expansion as it is called, is provided for on some scopes and is a valuable feature in television work. It is secured by incorporating an X amplifier and gain control. The circuit is arranged to give symmetrical expansion with respect to the centre of the screen. An X-shift control enables any required part of the trace to be brought on to the screen. A typical scope gives over ten diameters expansion, equivalent to a trace length of 50 cm or more.

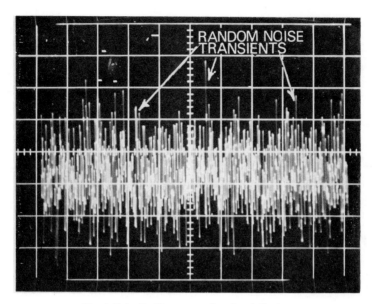

*Fig. 1.3(a)*. Oscilloscope noise-signal display.

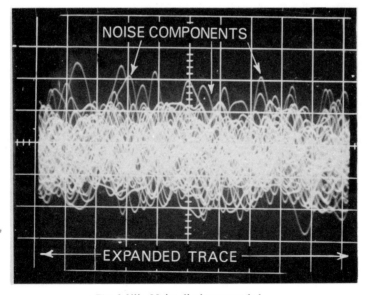

*Fig. 1.3(b)*. Noise display expanded.

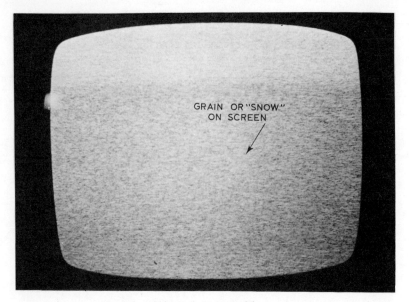

*Fig. 1.3(c).* Noise on television screen.

In Fig. 1.3 at (*a*) is shown an oscillogram (i.e. a picture or waveform taken direct from the screen of an oscilloscope) of 'noise' and at (*b*) a similar signal expanded. Note how the expansion widens the individual noise components and tends to result in confusion; it is noteworthy, however, that (*a*) is a photograph of a single sweep (single-shot display), while that at (*b*) is with the timebase in the repetitive mode. Noise on the screen of a television receiver is shown at (*c*).

As sweep speed is increased, so the trace reduces in brightness – at least in most service-type scopes. Devices are incorporated in expensive models to brighten the spot with increase in sweep velocity. As high sweep velocities are often required in television service work, the oscilloscope selected should produce a trace sufficiently bright to view in ordinary daylight at maximum sweep velocity. One should, of course, expect a reduction in brightness at full X expansion and at top sweep velocity.

**Bandwidth**

The true shape of a signal waveform is displayed only when the timebase is linear and when all the component frequencies making up the waveform are handled equally by the Y channel. This means the bandwidth of the Y channel is important.

The Y amplifier must have sufficient gain to provide the required sensi-

## INTRODUCTION TO THE OSCILLOSCOPE

tivity, and have it over the frequency range dictated by the application. Suppose, for instance, that an oscilloscope is required simply to monitor sine-wave signals between 50 Hz and 100 kHz; then the Y bandwidth need not exceed 100 kHz. For accurate amplitude measurement of waveforms, the Y channel response would need to be *flat* from 50 Hz to 100 kHz. There would be some response below 50 Hz and above 100 kHz, but the gain would be falling. On such a scope a sine wave at 150 kHz would almost certainly be displayed, but it would not be possible to measure its amplitude accurately by means of the Y attenuator. However, the sine wave would not be distorted and, if the effective gain of the amplifier at 150 kHz was known, the amplitude of the signal could be assessed.

It would be a different story with such a Y channel and a complex waveform. If a signal was a *square* wave with a repetition frequency of, say, 50 kHz, one might expect a Y channel with 100 kHz bandwidth to handle it without trouble. In fact, the trace would be distorted. This is because a square wave, in common with most complex waveforms, is given its characteristic shape by the mixing of many sine waves starting with one having the same fundamental repetition frequency as the square wave. A square wave contains sine-wave components corresponding to the fundamental frequency and odd-numbered harmonics up to large orders if the waveshape is to be preserved, corresponding to $Af_1 + Af_3/3 + Af_5/5 + Af_7/7$, etc., where $f$ is the frequency and $A$ the amplitude. Of course, there is no such thing as a perfect square wave (which would change from minimum to maximum amplitude and from maximum to minimum amplitude in zero time) but the point is this, that the repetition frequency and the frequency band occupied by a signal can be two different things.

In general, in dealing with anything other than one pure sine wave a Y amplifier should be capable of passing signals of frequency up to at least ten times the fundamental or repetition frequency without loss of amplification. Thus, to obtain a good display of a 50 kHz square wave the scope's Y channel should have a flat bandwidth up to at least 500 kHz. Considerable rounding of the corners of a square wave results from passing it through a channel of restricted bandwidth, as shown in Fig. 1.4(*a*). This is because the higher-order sine-wave harmonic components are either eliminated or attenuated.

Top repetition frequency encountered in television timebases is 15,625 Hz (the line frequency of the 625-line standard), so for displays of television timebase waveforms the bandwidth of the Y channel should be flat up to about 160 kHz at least. The response will not suddenly fall to zero at this frequency, of course, and there would almost certainly be useful gain up to 500 kHz or so.

It is sometimes instructive to obtain a display of components of the television picture signal. Some of these are of square-wave nature, or

7

SERVICING WITH THE OSCILLOSCOPE

transient pulses, and demand a bandwidth at least equal to that of the set's video amplifier (say up to 5 MHz). Oscilloscopes for colour television servicing require in general a wider Y bandwidth, not only to accommodate the subcarrier and video signals but also to ensure maximum definition with the least distortion of the fast-occurring pulse and transient types of signal. A Y amplifier bandwidth of, say, 10 MHz at the required Y sensitivity is often regarded as the minimum requirement. For audio applications a Y bandwidth well above the upper audio frequency is necessary when transient tests are made with square waves and signal pulses.

**Rise time**

How well an amplifier handles complex and transient types of signal depends on how quickly it can respond to sudden changes in amplitude of the input signal. This is the rise time, and for any active or passive transmission network it is usually defined as the time of rise of the output voltage from 10 to 90 per cent of its final value when a perfect step (i.e. input signal with zero rise time) is applied to the input. Most oscilloscopes are designed for minimal pulse distortion, and provided the pulse overshoot distortion does not exceed about 2 per cent, the following relationship between bandwidth $B$ and rise time $T_r$ obtains:

$$B = \frac{K}{T_r}$$

where $K$ is a constant governed by the response characteristic. When $K$ is characteristic of the so-called 'gaussian filter' whose frequency response is shaped like $\exp(-\omega^2)$ and impulse response like $\exp(-t^2)$ (i.e. when the $-3$ dB frequency corresponds approximately to half the $-12$ dB frequency), $K$ is taken as 0·35. The constant, however, can vary between about 0·3 and 0·5, depending on the precise nature of the roll-off characteristic.

$K = 0.35$ is generally true for Y amplifiers, which means that a 10 MHz oscilloscope is capable generally of a 35 ns rise time (1 ns $= 10^{-9}$ s). Note that the rise time is in seconds and the bandwidth in hertz (corresponding to the upper-frequency $-3$ dB point) so far as the expression just given is concerned.

Thus the rise time of an oscilloscope for colour television work should be around 50–20 ns. Oscilloscopes with less fast rise times and hence more restricted bandwidths are suitable for certain applications, but for transient observations the smaller the rise time the better.

The square-wave response in Fig. 1.4(a) is adjusted on the screen relative to the graticule so that the rise time can be measured. Here the sweep is 10 µs/div (horizontally) and the amplitude adjusted so that ten half divisions are embraced by the square wave. Thus it can be seen that the time between the 10 and 90 per cent amplitudes corresponds to about 4 μs, which is the rise time of this particular display.

This was taken by passing a very small rise-time square wave through an amplifier, and the oscilloscope used had a bandwidth of 25 MHz (i.e. 14 ns). Thus, owing to this and the fact that the rise time of the square wave (from the generator) was less than 100 ns, the display gives a fairly accurate assessment of the rise time of the amplifier under test. However, compensation needs to be applied when the signal rise time approaches that of the oscilloscope, and this is based on the expression

$$T_d = \sqrt{(T_s^2 + T_o^2)}$$

where $T_d$ is the displayed rise time, $T_s$ the signal rise time and $T_o$ the oscilloscope's rise time. Thus by rearrangement we get

$$T_s = \sqrt{(T_d^2 - T_o^2)}$$

Where more networks, etc., are connected in cascade, the rise time of the combination ($T_c$) is given by

$$T_c = \sqrt{(T_1^2 + T_2^2 + T_3^2, \text{etc.})}$$

where $T_1$, $T_2$, $T_3$ etc. are the individual rise times of the network.

## L.F. response

The low-frequency response of the Y channel is important for the accurate display of signal waveforms of low repetition frequency. Fig. 1.4(*b*) shows how a square-wave input may be distorted by a restricted low-frequency Y amplifier response. Normally, however, the majority of scopes have l.f. characteristics adequate to handle the field repetition frequency of 50 Hz without undue distortion.

A reasonable l.f. response is needed when the Y amplifier is brought into action in a wobbulator set-up for the visual alignment of tuned circuits and rejectors.

Some scopes have a Y bandwidth which depends on the sensitivity gain setting. At the highest sensitivity the bandwidth may be considerably less than at the lower settings. Instruments of this kind are usually suitable for television servicing, for the signal amplitude in the video channel, where a fairly wide Y bandwidth is required, is sufficiently high to apply to a Y channel working at reduced gain.

The Y amplifier must deliver sufficient output for full deflection without distortion. It is sometimes possible to overload a Y amplifier and get a display with flattened top, bottom or both (Fig. 1.5). Usually the design of the Y amplifier ensures plenty of reserve, distortion-free output even at full vertical deflection, but it is as well to check this.

## Sync facilities

Modern scopes have 'trigger sync' facilities, as distinct from the 'repetitive

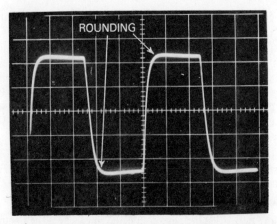

*Fig. 1.4(a)*. When the bandwidth of the amplifier or network is restricted, corner rounding of a square wave occurs due to the attenuation of the higher-order harmonic components. This display is arranged to show the measurement of rise time. The sweep is 10 µs/div, so the time between the 10 and 90 per cent amplitudes corresponds to about 4 µs, which is the rise time. The repetition frequency of this display is 20 kHz.

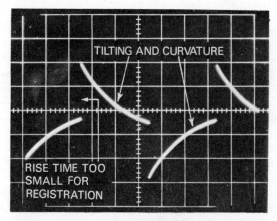

*Fig. 1.4(b)*. Square-wave distortion produced by low-frequency differentiation, indicating a too early bass roll-off for the square-wave frequency. The slight curvature of the top and bottom parts of the waveform signifies phase distortion.

# INTRODUCTION TO THE OSCILLOSCOPE

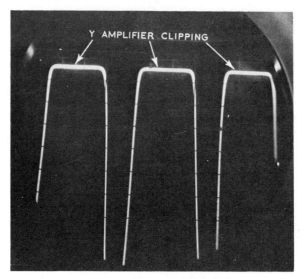

*Fig. 1.5.* Clipping in the Y amplifier.

sync' arrangement of earlier models. Some service-type instruments are switchable to repetitive or trigger sync on positive or negative-going waveforms, obtained either from inside the instrument or externally and then fed in through the 'sync terminal'. Those whose designs cater particularly for television applications have a 'television field' sync input, allowing the field sync pulses of the set to be used to 'lock' the scope's waveform display.

With repetitive sync some of the test signal is applied to the timebase to lock the display on the screen. This sync signal is either coupled to the 'sync terminal' direct or is switched to the timebase from the Y channel inside the scope. Synchronisation is effected by turning down the sync amplitude and adjusting the fine sweep control until the display is almost stationary. Then, optimum lock is achieved by turning up the sync amplitude.

Too much sync signal should be avoided as it can distort the display. The sync may need to be reset each time the test signal frequency or the sweep frequency is altered.

Trigger sync has a distinct advantage over this. Here it is the test signal which actually instigates the forward trace of the timebase. When the test signal is applied the timebase is fired by the signal on its positive or negative half-cycle (as required) independently of the signal characteristics. Any change in frequency fails to affect the locking of the display. All that happens is that an increase in signal frequency increases the number of complete cycles of waveform displayed on the screen. This makes life much easier than having to readjust each time, as required with the repetitive arrangement.

## Other facilities

*Delayed sweep*

More advanced instruments include such refinements as delayed sweep, mixed sweep and switched sweep. With the former, which is a common facility, two timebase generators are utilised, but with the display being, in one mode, of the second generator. The first is called a *delaying generator* and the second the *delayed generator*. The ramp produced by the delaying generator is instigated by a trigger pulse and the ramp continues until it reaches the level (*comparator level*) established by the setting of the delayed generator. When the delayed generator is in the *auto* mode it operates as soon as the delaying sweep reaches the comparator level, the first generator thus being a *time-delay generator*. The trigger-delay mode is sometimes termed *arming* because it readies the trigger-delay generator to operate on the next trigger pulse.

Delayed sweep offers improved accuracy and definition of time-interval measurements, and one oscilloscope used by the author has facilities such that the start of the sweep may be delayed by between 500 μs and 50 ms after the triggering signal by the selection of appropriate delay buttons and by the use of coarse and fine delay controls.

*Single shot facility*

This is generally a button which when depressed 'fires' the timebase so that it makes one complete sweep only. Either internal or external triggering may generally be used. To 'arm' the timebase for 'firing' there is a second push-button which when depressed causes a neon to glow. After the sweep has been concluded the neon extinguishes.

This facility assists in viewing or photographing a non-repetitive signal or random event. The noise display in Fig. 1.3(*a*) was photographed in this manner.

*Y amplifier delay*

Oscilloscopes for professional use, particularly those of extended Y bandwidth, commonly include delay lines in the vertical amplifiers. The idea is to delay the Y signal so that it is kept from the deflection plates of the display tube until the horizontal sweep commences. If this is not done the leading edge of a fast-rise pulse fails to be displayed. The Y delay of the author's Telequipment D53 scope, using JD vertical amplifiers, is 200 ns.

*Plug-in amplifiers, and differential amplifiers*

Some oscilloscopes, such as the D53 just mentioned, are designed for plug-in vertical amplifiers. Amplifiers of various gains and bandwidths, with or without differential inputs, can be accommodated to suit best the job in hand.

Differential amplifiers are particularly advantageous for Y amplification since they have the capacity to reject unwanted spurious signals, whether radiated or inductively or capacitively coupled. Such amplifiers also exhibit less drift and hence improved stability over a wide range of operating conditions.

They also facilitate accurate comparative measurements, such that when a reference signal is applied, say, to the negative input and the test signal to the positive input the display constitutes the difference between the two signals.

Very advanced instruments also incorporate plug-in timebase units. In this way, then, the scope can be purchased as a basic main frame and the various units plugged in to suit the test or measurement requirements, thereby providing enhanced flexibility in changing the instrument's measurement capability.

*Dual trace instruments*

For many tests and measurements the dual or even single trace scope is adequate, though multiple trace instruments are required for research and other more detailed applications.

The dual trace instrument is capable of providing two simultaneous displays, each one via its own Y amplifier, with, in general, the timebase being common to both displays, but with the trigger or sync being switchable to either.

The two traces are obtained by dual or split beam cathode-ray tubes or by the use of beam switching, where a common beam is switched rapidly between the two displays.

*Storage oscilloscope*

Single-shot events and signals of very low repetition rates are often difficult to view on an ordinary oscilloscope owing to the relatively short time of phosphor decay. A storage oscilloscope solves these problems since the display is held on the screen for leisure viewing and analysis. After such study or photographing the display can be extinguished.

**Choosing a scope**

We have now covered the fundamental features of a scope and with these in mind there should not be great difficulty in selecting a scope to satisfy the economics of the service department. It may well be that the selected instrument will have to aid in the servicing of other equipment, such as audio amplifiers and tape-recorders. This may modify requirements a little but a scope with the specification outlined here will cater for a wide range of applications.

Screen size is important, and the general rule is the larger the screen the better, but the size is often dictated by price. The display areas of modern, professional instruments are often rectangular, as shown by the oscillogram in Fig. 1.4(a), for example. Here the graticule is divided into 1 cm squares, providing an active screen area of $10 \times 8$ cm, which is a useful size. Smaller screen areas, however, are suitable for service-department applications.

For the majority of servicing applications the single-beam scope is adequate but the dual-beam or split-beam instrument does allow the simultaneous display of two Y signals. This can be handy for assessing the distortion produced by an amplifier, for example. The input signal is applied to one Y channel and the output signal to the other and, after equalizing the amplitudes of the two displays, the Y shift controls can be adjusted so that the displays fall one on the other. Any distortion of the output signal relative to the input signal is then easily seen.

Another application of a dual-beam scope is illustrated in Fig. 1.6. Here to one Y input was applied the audio modulating an r.f. carrier wave while to the other input was fed the modulated signal. The oscillogram shows clearly the presence of modulation distortion. To be correct the envelope of the modulated signal should exactly coincide with the modulation waveform.

Beam switching devices are available for use with single-beam scopes,

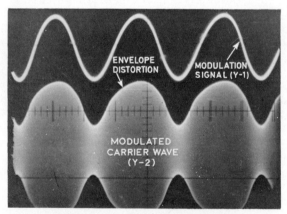

*Fig. 1.6.* Sine wave modulation signal on Y-1 and modulated carrier wave signal on Y-2. Note the envelope distortion.

allowing them to take two Y inputs and give simultaneous displays. They have certain bandwidth restrictions, however.

After deciding on the basic requirements, it is prudent to investigate how a particular instrument matches up to its specification, but it must be remembered that the lower the price the greater will be the limitations, relative to a more expensive instrument of similar facilities. The price, of course, also reflects the facilities offered, and on a rough price scale, starting at the lower end, general-purpose scopes can be classified as follows.

1. Single trace of bandwidth up to 10 MHz and maximum sensitivity of 10–100 mV/cm with 7·5–12·5 cm tube operating at 2–3 kV.
2. Dual trace with parameters as 1 above.
3. Dual trace of 20 MHz bandwidth and maximum sensitivity of 5–10 mV/cm with 12·5 cm tube operating at 4 kV and single timebase.
4. Dual trace of 25–40 MHz bandwidth and maximum sensitivity of 5–10 mV/cm with 12·5 cm tube operating at 10 kV and single or dual timebase.
5. Dual trace of bandwidth greater than 40 MHz and maximum sensitivity of 5–10 mV/cm with 12·5 or 16·5 cm tube (rectangular display area) operating up to 20 kV and dual timebase.

Clearly, these classifications are by no means exhaustive and should be taken essentially as a rough guide.

The e.h.t. voltage or tube accelerating voltage gives some indication of the relative brightness to be expected, though there are other internal and external factors that can affect this. Other useful points to note are:

1. *Brightness control adjustment.* It should be possible to extinguish the trace on the one hand and outweigh the retrace blanking on the other.
2. *Focus.* Minimal defocusing should occur when the spot is shifted from centre to each corner of the screen in turn, and when brightness is adjusted.
3. *Trace distortion.* With the Y input shorted and the timebase running the resulting horizontal line should exhibit minimal pincushion and barrel distortion when shifted in turn to line up with a graticule line at the top and bottom of the screen. The distortion should also be minimal when a vertical line, produced by switching the timebase off and applying a sine-wave signal to the Y input, is shifted from the left to the right of the screen. Some instruments include a trace rotation preset for aligning the trace to a graticule line. See Fig. 1.7(*a*).
4. *Y amplifier response.* As has already been mentioned, the Y bandwidth may be less on the higher sensitivity ranges than on the lower ones. This parameter may govern the choice when a scope is likely to be employed for the measurement of fast-rise pulses in low-level circuits. The pulse characteristics can be examined by applying a fast-rise pulse to the Y

input and checking the display for overshoot, undershoot or ringing (the effective rise time of the pulse, of course, should be smaller than the scope's rise time). Medium-frequency compensation can be examined by viewing a square wave over the range of about 100 Hz–10 kHz at all sensitivities. There should be no overshoot or ringing (assuming that the source signal is free from such effects!) and the tops and bottoms of the waves should be flat (i.e. minimal tilt). Similar tests should be made with a compensated divider probe if such an accessory is to be used with the instrument (see the next section).
5. *Inter-channel crosstalk*. With a 10 kHz square-wave signal applied to the input of one channel, the trace of the other channel should be examined for signs of breakthrough, both with its input shorted and open-circuit. Checks should be made at all sensitivities, also when an overload signal is applied to the speaking channel.
6. *Sweep linearity*. With a sine-wave signal applied to the Y input, the sweep time should be adjusted for the display of an integral number of cycles per division of sweep. Non-linearity is revealed by checking the spacing of the cycles against the graticule. Check in the same way with X expansion applied. Non-linearity occurs mostly over the first and final 10 per cent of a sweep, particularly at the higher sweep velocities. See Fig. 1.7(*b*).
7. *Sync and triggering*. Checks should be made of sync and triggering (most modern instruments are essentially triggered by the signal rather than synchronised to it, and some include a 'bright line' arrangement whereby the sweep is free-running in the absence of trigger signals) on all modes and at all sensitivities and frequencies. In particular, check the effectiveness of triggering on a high-frequency Y input signal. Breakthrough interference from the triggering circuits is often revealed by the presence of small spurious signals along the display that move as the trigger level control is adjusted. Instruments designed for television servicing applications commonly include line and field triggering facilities. These are basically simple high-pass filtering for the line signals and low-pass filtering for the field signals.
8. *Hum pick up*. Short the Y input and set the timebase to about 20 ms/div and obtain a slow drift of hum on the trace by adjusting the fine sweep control. Hum on the trace should be negligible under this condition. Also retard brightness and check for 50 Hz brightness modulation on the trace.

**Compensated voltage-divider probe**

One way of connecting Y signal to a scope is through a simple length of wire, but this suffers from hum and spurious signal pick-up. An alternative is screened cable. This is often satisfactory for relatively low-frequency signals and pulses of not too small rise time, but for higher frequencies and

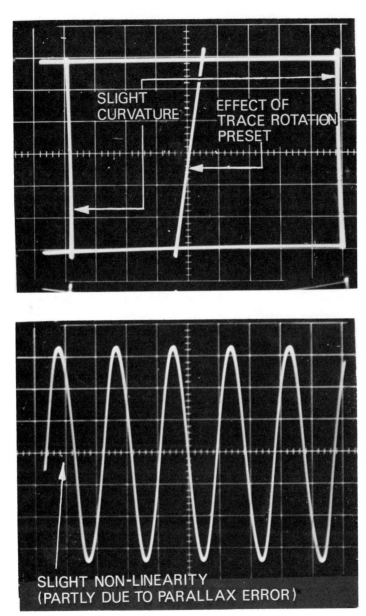

*Fig. 1.7.* (*a*) Check for trace distortion (see text). (*b*) Check for sweep linearity (see text).

faster pulses the signals can be attenuated or distorted by the input shunt capacitance of the screened cable.

The compensated voltage-divider probe combats this sort of problem by presenting to the circuit under test a higher load resistance and lower shunt capacitance than the Y input connected through screened cable, but at the expense of input attenuation, which needs to be taken into account when signal amplitudes are measured against the graticule. Probes are available with 10:1 and 50:1 voltage attenuation ratios, corresponding respectively to 20 dB and about 34 dB. Thus when a signal amplitude is being measured with the scope's input attenuator the attenuation of the probe (in decibels or direct ratio) must be used to correct the attenuator/graticule value.

Active probes are also available with a 1:1 gain ratio, but these tend to apply dynamic range limitations on the input (about $\pm 0.5$ V to $\pm 5$ V). The active device is often a field effect transistor. The passive 10:1 probe is the most popular, the circuit of which is given in Fig. 1.8. The compensating trimmer is adjusted for a non-tilted square-wave signal. (Note: some scopes feature a probe test point that delivers a fast-rise positive-going pulse of about 6 V p–p, and the probe is adjusted for a square corner on the displayed pulse when the tip of the probe is held against the test point.) Care has to be taken to avoid over-compensation, which is equally as bad as under-compensation.

The rise time capability of the probe is generally specified, so the total rise time from probe tip to the c.r.t. is obtained from

$$T_t = \sqrt{(T^2_{\text{probe}} + T^2_{\text{scope}})}$$

where $T_t$ is the overall or total rise time, $T_{\text{probe}}$ the rise time of the probe and $T_{\text{scope}}$ the rise time of the oscilloscope.

A probe of this type is often specified for tests and adjustments in colour television receivers.

### Oscillograms

It seems desirable at this point to add a note about the oscillogram photography used in this book. In general, a photograph gives very much the same 'picture' of the oscilloscope screen as the eye sees; but differences do occur, and their nature and the reasons for them should be appreciated.

Ideally, the camera shutter should be synchronised to the scope so that the shutter opens as the spot starts from the left and closes during the retrace. That way the film would be exposed to one trace only. This method is sometimes employed (especially on single-sweep displays), the sweep being 'triggered' rather than under the control of a timebase.

For oscillograms like some of those used in this book, however, the camera shutter remains open during a number of scope-synchronised sweeps. Certain minor problems result from the traces not falling exactly on top of each other. First, small changes in signal conditions brought about by

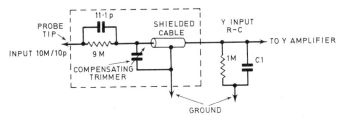

*Fig. 1.8.* Circuit diagram of compensated 10:1 voltage attenuation ratio voltage-divider low-capacitance probe.

random noise, interference, thermal effects, spurious signal pick-up and so forth, and small changes in mains voltage (particularly at sweep velocities approaching 1 μs/cm), slightly displace traces in whole or part relative to each other, so thickening the trace in whole or part on the photograph. Secondly the horizontal parts of traces tend to be thicker because the film is 'exposed' to the spot for a longer time during horizontal movements than during the faster vertical movements. Thirdly when a waveform is occurring much more rapidly than the sweep, the separate cycles diffuse to give an area of illumination. This effect is illustrated in Fig. 1.9. Here the higher frequency line signals form the area of illumination. The 'envelope' of this display is as shown a little shaped by hum (50 Hz) pick-up. On the set this would show as a hum band (not very intense) occurring once on each field.

The field pulse (Fig. 1.9) may seem a little curious but it must be remembered that the field sync signals comprise a series of pulses (each one about

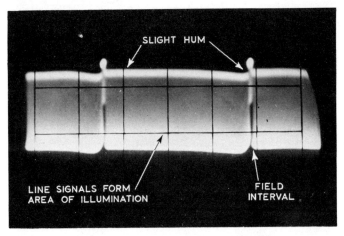

*Fig. 1.9.* Line signals occurring much more rapidly than the sweep, causing diffused area of illumination.

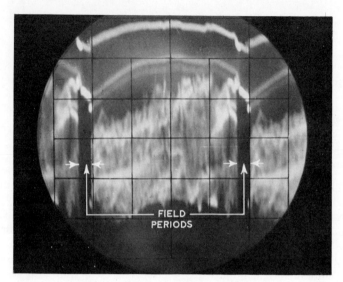

*Fig. 1.10.* The field periods in greater detail.

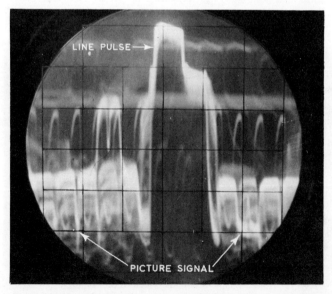

*Fig. 1.11.* Here the scope was synchronised to the line sync pulse.

## INTRODUCTION TO THE OSCILLOSCOPE

40 us on the 405-line standard) dropping from black to blacker-than-black. The gaps in the display in Fig. 1.9 represent the period of this series of pulses, while the odd-looking little blobs indicate a change in the video operating conditions when the picture signal ceases. The action here will be detailed in subsequent oscillograms, and the various component parts will be identified.

The field period is revealed in greater detail in Fig. 1.10. Since it is the integrated pulse derived from this period that 'fires' the field generator, a back-coupled pulse from this generator is sometimes seen on this sort of waveform, depending upon where the signal is taken from and the design of the field sync coupling from the sync separator.

Fig. 1.11 was obtained by synchronising the scope to the line sync pulse. Relative to this, the picture signal is for ever changing and this produces subsidiary traces on the display, as the oscillogram shows. Just how well such traces are defined depends on the Y bandwidth which itself – like the television's video amplifier – governs the display definition.

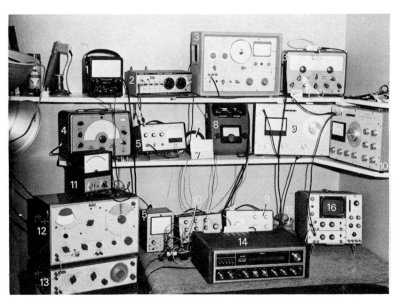

*Fig. 1.12.* Some of the test equipment used by the author for the measurement of hi-fi parameters and for producing oscilloscope displays. 1: Valve voltmeter. 2: Decade attenuator. 3: F.M./A.M. generators. 4: A.M. generator. 5: Audio millivoltmeters. 6: Mains voltage monitor. 7: Audio meter. 8: Electronic test set. 9: Low-distortion a.f. oscillators. 10: Distortion factor meter. 11: Multirange meter. 12: Wave analyser. 13: Notch filter. 14: Equipment under test. 15: Varactor. 16: Dual-beam oscilloscope.

21

## SERVICING WITH THE OSCILLOSCOPE

Some of the oscillograms contained in this second edition were obtained with a Polaroid oscilloscope camera using black and white Type 107 film packs (giving $3\frac{1}{4} \times 4\frac{1}{4}$ inch prints). Both repetitive and single shot techniques have been adopted, the former at 1/30 s shutter speed and f8 aperture and the latter at f11 aperture with the shutter open, the exposure being made with the single shot facility on the oscilloscope.

For oscilloscope photography the scope's graticule should be of the illuminated type and it helps to secure a desirable contrast between the trace and the background graticule if the illumination is adjustable, as it is on some of the author's oscilloscopes.

It also helps if the graticule can be interchanged for different measurement requirements. For example, the sweep-frequency displays in this book are based on a background produced by a graticule calibrated in frequency horizontally and in decibels vertically.

A 35 mm reflex camera with close-up lenses was used for the off-screen television displays. If possible, the shutter should be synchronised to the field sync pulses, but failing this a reasonable display can be photographed by arranging for the shutter to remain open for two interlaced fields, constituting one complete picture or frame. If only one and a bit fields are photographed a diagonal line appears across the picture, depending on the type of shutter used in the camera. Film used was Ilford FP4, the shutter speed 1/25 s and the aperture setting f2.8, with the brightness and contrast of the receiver adjusted for the best results.

# 2: APPLYING THE OSCILLOSCOPE

THIS chapter will explain the basic television and audio applications of the oscilloscope, how the controls are adjusted and how the screen displays are interpreted.

**Television applications**

When probing with a voltmeter we tend to forget that a television set is a series of circuits arranged to produce voltage and current signals (using that word in the broad sense) of accurate timing and specific shapes and amplitudes. Voltages across the various elements, and currents in them, as shown by a d.c. test set, tell very little about the pulsating, live nature of the signals.

Simple meter tests indicate that a circuit is *probably* capable of handling signals according to design. Actually, it is possible for a set to stop working and yet the d.c. conditions to differ only slightly if at all from those normally present. D.C. conditions set the stage, as it were, for the signals. Clearly, if the d.c. conditions differ substantially from normal one can be pretty sure that the signal-handling is not as it should be. But 'd.c. conditions normal' is not conclusive indication that signal conditions are also normal.

The oscilloscope is the only instrument that can show what is happening to the amplitude, frequency and shape of signals. Therefore it has considerable diagnostic potential. In fact, logically, it should be used for all preliminary tests. Afterwards, if necessary, the multirange testmeter would pinpoint the faulty item.

The scope is not as widely employed as it should be for several reasons. One is that technicians often consider it too much trouble to set up so that it is brought out only as a last resort. Another reason is that many service departments, although boasting a scope, possess a vintage instrument not suitable for television or hi-fi work. Still another reason, probably the most important, is that skill in scope diagnosis comes with practice and, since the instrument may be more often on than off the shelf, this is not obtained.

Knowing what to expect is important in scope work, and with the greater complexity of colour television receivers compared with other equipment, more manufacturers are including sketched or photographed wave-forms at various points on their circuits.

A good exercise is to hide away the multirange meter for a few weeks and use the scope for all tests – d.c., a.c. and signal.

Now to consider how to apply the scope for various tests and what sort of waveforms to expect. The scope offers distinct advantages when checking in the timebases. For instance, lack of e.h.t. can be caused by trouble in the e.h.t. rectifier, the line output stage or the line generator stage. Without making any connection but one, the scope can be used to tell whether or not the line generator and output stage are working. All that is necessary is to clip the earth lead of the scope to the metalwork (h.t. negative) of the set and place the Y-input lead near the line output transformer. The field radiated by the timebase is so strong that good vertical deflection on the screen of the scope will be given with the Y attenuator set to about 30 V/cm, two pulses being displayed when the sweep is set to about 30 μs/cm. A waveform so obtained (Fig. 2.1) shows the peak amplitude of the pulses picked up on the lead, which is of the order 3 × 30 or 90 V, as the pulses occupy three 1 cm squares on the graticule.

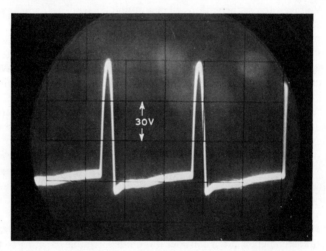

*Fig. 2.1.* Line pulses picked up on the oscilloscope Y lead. On 625 lines, the time between pulses is about 64 μs and the time period of the pulses about 10–15 μs.

A set exhibiting lack of e.h.t. yet giving a display as in Fig. 2.1 would probably have a faulty e.h.t. rectifier. Anyway, one could say definitely that the line generator and the line output stage were working correctly regarding both amplitude and frequency.

The energy in a correctly-operating line output stage is considerable and

## APPLYING THE OSCILLOSCOPE

care should be taken to avoid letting scope leads come into direct contact with a terminal on the line output transformer. As we have seen, sufficient pick-up of signal is possible by random coupling.

Care must also be taken, of course, to ensure that the chassis is adequately isolated from the mains supply. It pays to use a 1:1 ratio mains isolating transformer where possible. The alternative is to make sure that the chassis is in connection with mains neutral, for full mains potential on the scope earth circuit could damage the instrument; and then there is also personal danger from electric shock to be seriously considered! Some colour receivers have a power supply circuit designed in such a way that the chassis is at half mains potential whichever way round the mains is connected. Transformer isolation is essential with such designs.

The transformer must be able to handle the full power of the receiver without core saturation or without seriously distorting the mains supply waveform.

**Random pick-up**

Random pick-up of line signal can interfere with waveform tests made in other, lower-level stages. For this reason it may be desirable to mute the line timebase when checking field, sync and other signals.

Caution must be observed when undertaking such a task. On no account should the line output valve be left devoid of line drive, as this would burn up the valve and not do the line output transformer much good either. One can mute the stage by removing the line output valve and plugging in a dummy valve base fitted with a wire-wound resistor of value equal to that of a hot heater. This trick maintains heater chain continuity and prevents other heaters from being over-run. A PL81 (for example) has a 21·5 V 0·3 A heater. Ohm's Law says $R = E/I$, where $E$ is the heater voltage and $I$ the heater current in amperes. So we have 215/3, which works out a little over 71 ohms. Wattage is 21·5 × 0·3, or 6·4. In practice, a 70 ohm 10 W resistor would be used.

Similarly, it may be desirable to mute the line generator or the field timebase (generator or amplifier). So it is a good idea to make up and keep at hand a selection of popular-type valve bases with suitable-value resistors across the heater pins. Care should be taken, of course, to see that there are no internal connections between other pins.

Alternatively, the screen-grid feed to the line output valve can be disconnected and the h.t. loaded accordingly.

The field timebase may have to be muted when checking for breakthrough of line pulses into the field circuits, an effect that can severely disturb the set's interlace performance.

Weak field sync can be diagnosed with a scope by checking the shape and amplitude of the field sync pulses leaving the sync separator stage and

## SERVICING WITH THE OSCILLOSCOPE

following them through to the field generator; but they cannot always be seen with the field generator running. The same applies to the line sync pulses in the line generator circuits.

Some manufacturers' service manuals illustrate the basic waveforms to be expected at vantage points in correctly working sections; but these are not a great deal of use unless their peak-to-peak amplitudes are indicated. It is possible, for instance, for a scope to show line drive at the control grid of the line output valve and yet, because of trouble in the output valve or stage, for the amplitude of the signal to be less than it should. This is because an output stage fault can sometimes alter the loading on the line generator, and thus alter the drive amplitude. Unless the correct drive amplitude is known, therefore, the diagnostic value of the scope is impaired. Nevertheless, with experience one can tell when the signal amplitudes at various sections are abnormally high or low.

In some circuits or service manuals the signal at the screen grid instead of at the control grid of the line output valve is shown. In this case, if it is found that the line drive from the generator is the correct shape and amplitude but the waveform at the screen grid of the output valve deviates from that shown on the circuit, then there is little doubt that the fault lies in the output stage – a fact worth knowing, especially when the line output transformer is under suspicion!

Circuit waveforms are not meant to show a lot of detail; they are meant, essentially, to check the basic operation of circuits. When the waveforms are actually produced on the scope they may not be recognised immediately. Indeed, they can be grossly altered in shape by the setting of the scope controls. For most television waveforms, the sweep can be set to about 3 ms/cm for field circuits and 10 μs/cm for line circuits. By the use of the continuously variable sweep control, these settings will give one, two or three complete waveforms on the screen.

Receivers with transistorised timebases cannot be so conveniently muted as those using valves. Reference should be made to the manufacturer's service manual, because endeavouring to suppress timebase action by electrode shorting or potential removing could well result in changes in d.c. conditions throughout the stage and in some cases destroy one or more transistors – so beware.

### High- and low-level inputs

In most normal tests the ordinary Y attenuator will permit adjustment within the peak-to-peak range of the waveform. But sometimes an external attenuator will be required between the Y input and the test circuit to avoid overloading the input stage of the amplifier. At the anodes of the field and line output stages, for example, the peak-to-peak voltages are often well in

## APPLYING THE OSCILLOSCOPE

excess of the scope's capabilities. The attenuator should be resistive and have a high input impedance and low input capacitance.

Low-level signals are best fed to the Y input through screened cable to avoid pick up of hum, but it must be remembered that waveform distortion is likely if the input circuit offers too great a capacitance.

**Nature of signal**

Many scopes have a switch marked 'a.c./d.c.'. In the a.c. position a capacitor is switched in series with the Y input and in the d.c. position there is a direct connection (through direct-coupled Y amplifier) to the Y plates of the cathode-ray tube. In the former position, the scope will fail to indicate d.c. voltage but the series capacitance is of such value that good response is possible down to very low a.c. frequencies (see the particular specification). In the latter position, the spot will shift vertically to indicate the d.c. voltage as well as deflect up and down on a.c. A waveform with an a.c. peak-to-peak value of, say, 10 V will give the same amount of deflection as a d.c. potential of 10 V.

The peak-to-peak value of a sine wave is twice its peak value (see Fig. 2.2(a)) and the root-mean-square (r.m.s.) value of a sine wave is 0·707 of its peak value. Thus, a sine wave with an amplitude of 10 V peak-to-peak has a peak amplitude of 5 V and r.m.s. value of 3·535 V.

This means that the 'heating effect' value of a 10 V peak-to-peak sine wave is equal to only 3·5 V d.c. although the *deflection* on the scope is the same for 10 V d.c. and 10 V peak-to-peak. This, of course, is because a cathode-ray tube responds to peak voltage whereas a rectifier and moving-coil voltmeter (for instance) is calibrated in r.m.s. values. It is possible for a scope to be *calibrated* similarly.

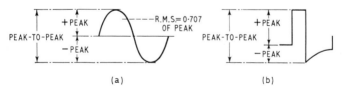

*Fig. 2.2.* Comparison of sine wave and non-sine wave peak-to-peak values. (a) The peak-to-peak value of a sine wave is twice its peak value and the root-mean-square value is 0·707 of its peak value. (b) The peak and peak-to-peak values of a non-sine wave.

The foregoing applies to sine waves. Actually few signals in a television set are sine waves. Many are of a pulse nature. Sine waves are symmetrical, having equal positive and negative peaks, but pulse waveforms are essentially asymmetrical, see Fig. 2.2(b). Nevertheless, the sum of the two peaks still

SERVICING WITH THE OSCILLOSCOPE

is the peak-to-peak value, and it is this value in which we are mostly interested.

Fig. 2.3 shows the luminance emitter follower of a colour receiver with the input and output waveforms. The first in each case is with the scope sweeping to suit the line signal and the second in each case with the sweep velocity reduced to show the field signal. The line signal occupies a time of about 64 μs and the field signal about 20 ms, so the timebase must be set accordingly, depending on how many complete line or field signals it is proposed to display. It will be seen that the line signal contains colour bar information (see Chapter 8) and also shows the line pulse and bursts. Since the stage is in common-collector mode (i.e. emitter follower, with the signal emanating from the emitter) there is no phase reversal of the output signal with respect to the input signal. Fig. 2.4 shows how a field signal actually appears on the screen of a scope. Fig. 2.5 has trace expansion applied to secure greater detail. These displays are inverted relative to those in Fig. 2.3.

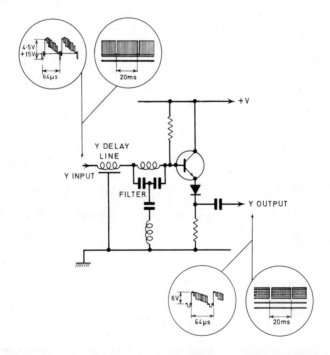

*Fig. 2.3.* Luminance emitter follower stage of colour receiver, showing input and output waveforms at both line and field frequencies. The signal represents the standard colour bars (also see Chapter 8).

# APPLYING THE OSCILLOSCOPE

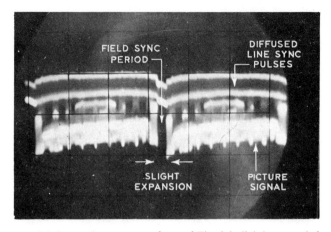

*Fig. 2.4.* Inverted output waveform of Fig. 2.3 slightly expanded and as actually seen on the scope.

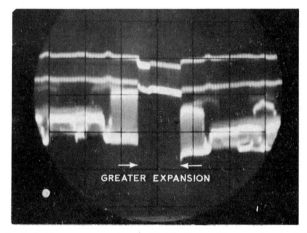

*Fig. 2.5.* Greater expansion applied to the oscillogram in Fig. 2.4.

Here we are considering mainly amplitude; later chapters consider similar waveforms from other aspects.

Fig. 2.6 shows a single line pulse display of a monochrome signal with expansion to give some idea of the composition of the signal either side of the positive-going line pulse. Note the rounding due to a falling h.f. response.

29

SERVICING WITH THE OSCILLOSCOPE

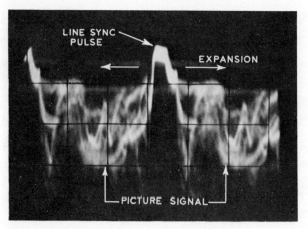

*Fig. 2.6.* Single line pulse display with expansion.

**Audio applications**

The oscilloscope and waveform generator are two items of equipment essential for serious servicing work on quality audio amplifiers. It is possible, of course, to trace the cause of total failure of an amplifier with nothing more elaborate than a multirange testmeter, coupled with experience and a knowledge of the equipment under test. But, mostly, hi-fi amplifiers suffer from troubles like lack of full output power, distortion, instability, ringing and hum, and the ordinary testmeter is no great help in the diagnosis of dynamic faults of this kind.

By the injection of a signal into the amplifier any deviation from the known parameters of the input waveform can be readily observed anywhere in the circuit on an oscilloscope. This is dynamic testing. The audio generator should produce a pure sine wave output adjustable from zero up to about 10 V r.m.s. over the frequency range 20 Hz to at least 100 kHz. Preferably, it should also produce square waves of the same voltage and repetition frequency. The oscilloscope should have a Y bandwidth of at least 1 MHz and a Y sensitivity of no less than 100 mV/cm. The X timebase should be adjustable over the range 100 ms/cm to 1 μs/cm, preferably with expansion. Sync should be repetitive or triggered (preferably both).

A double-beam instrument is useful since the output waveform can then be compared directly with the input waveform but the author has not found such an instrument to be essential in ordinary service work. A good single-beam model is better than an old-style double-beam instrument. An electronic beam switch can be used to convert a single trace model into a dual trace display.

## APPLYING THE OSCILLOSCOPE

Generator and oscilloscope allow the maximum power output of an amplifier to be checked. A sine-wave signal is applied to a 'flat' input and the level adjusted to avoid overloading the early stages. Instead of a speaker, the output is developed across a resistor of a value matching the output impedance and a rating to take the maximum watts. The resistor should be non-inductive and possess a linear resistive characteristic and should be of suitable rating to handle the full average sine-wave power output of the amplifier without overheating. A heat-sink-mounted resistor may be required for amplifiers of large power.

The Y input of the oscilloscope is connected across this load and the attenuator adjusted to give about a half to two-thirds vertical deflection. The input frequency should be 1,000 Hz, meaning that the sweep should be set to about 500 μs/cm to give two full waveform displays, depending on the setting of the 'fine' sweep control.

With the tone controls 'flat', the input should be advanced until waveform clipping is observed. This should be symmetrical. That is, the clipping should be about equal on the tips of each half-cycle as in Fig. 2.7. The input should now be reduced until the waveform becomes pure, and the peak-to-peak amplitude then measured on the graticule. In Fig. 2.7, for instance, the p–p amplitude is about 5 cm. As the Y input is set to 10 V/cm, the p–p amplitude is 50 V. The peak amplitude is half this value (25 V) and the r.m.s. value (required for the power measurement) is 0·707 of this or a little over 17·5 V.

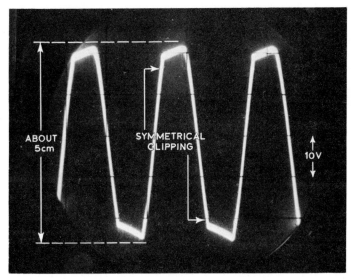

*Fig. 2.7.* Display of symmetrical clipping.

## SERVICING WITH THE OSCILLOSCOPE

To find the audio power, we square the r.m.s. voltage and divide by the value of load resistor. The square of 17·5 is 306·25 and, if the load is 15 ohms, the power works out to about 20 W.

Should clipping commence well before the rated full power is reached, tests can be made to determine whether the trouble lies in the drive waveform or in the output stage. First, however, remember that transistor amplifiers may have 'music power' rating. This differs from continuous-wave average power. On running such an amplifier hard on sine wave, the output transistors may overheat and fail. For thorough checking it is essential to know the average power of an amplifier when its rating is given in music power. The average power will be somewhat below the music power.

This is but one audio application of the scope. Others, with oscillograms, are described in Chapter 10, while Chapter 7 reveals how the versatile scope can give visual circuit realignment in radio and television sets.

# 3: VIDEO WAVEFORMS

A VIDEO waveform as considered here contains not only the picture signal but also the line and field sync pulses. These components, and sometimes spurious signals, appear at the vision detector load for conveyance to the video amplifier or output stage. What the waveform looks like on the scope depends on the settings of the scope controls, on how well the vision channel of the set is aligned, and on whether any of the signal is lost or distorted during transference from set to the scope's Y input. The most convenient place to pick up the video signal (prior to the video amplifier) is at the output of the video detector.

The connection to the scope must be low capacitance. If a special low-capacitance test probe is not used, then the 'live' Y lead should be kept well clear of other leads and metal parts. It may be necessary to use screened cable to avoid hum injection, but as such cable has appreciable capacitance between conductor and braid it is important to keep it as short as possible. Whether or not hum will be troublesome is determined by the signal/hum ratio at the scope Y input. If the signal is of high amplitude, average hum pick up on the Y lead – even if unscreened – will not materially affect the display. However, if the signal is weak, calling for a high Y gain on the scope, hum will almost certainly affect the display unless reasonable precautions are taken.

The square-wave nature of some components of the video signal indicates the presence of high (harmonic) frequencies, as explained in Chapter 1. Lack of Y bandwidth will attenuate these and distort the waveform. A 'bad' display will also result with a set having misaligned vision i.f. channel or from one in which the vision detector is faulty, the detector load resistor increasing in value and restricting the bandwidth. Loss of the higher frequency harmonics in the connecting leads – due to too much shunt capacitance – similarly affects the display. Even a very good display is rarely up to textbook standard!

**Off-air waveforms**

For a video test, the input to the receiver can be from a video signal generator or an 'off-the-air' test signal. The latter is generally better as it is

SERVICING WITH THE OSCILLOSCOPE

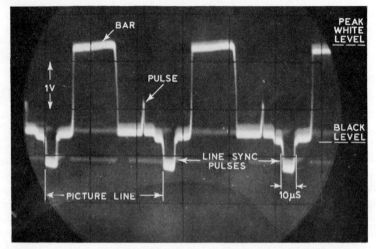

*Fig. 3.1.* Video test waveform at the input to the video amplifier.

under the control of the transmitting authority – and it costs nothing.

Figure 3.1 shows a waveform picked up at the input to the video amplifier. Here we have three picture-lines of signal. From the first sync pulse (left of picture) the signal remains for a short period at black level. It then rises quickly to white (towards full amplitude on the 405-line standard) and holds this for about half a line. It falls to black level again but, before the line is completed, rises sharply giving a pulse half-way along the black period. Then the signal falls to 'blacker-than-black' to give the second line sync pulse.

This is a waveform representation of the test transmission, seen from some local stations, giving a black vertical band on the left of the screen, then a wide white vertical band followed by a black band in which the white, very narrow vertical line occurs (due to the pulse on the waveform).

It will be noticed that the horizontal parts of the photographed waveform are considerably thicker than the vertical parts. This is due to some hum pick-up and slight bouncing of the waveform while the photograph was taken. The scope's timebase coarse control was set at about 20 μs/cm and the variable sweep control adjusted to lock three lines. The Y amplitude was set to 1 V/cm.

Test signals of this kind are probably the best for scope displays, for reasonably good response characteristics of i.f. strip and vision detector are clearly revealed by the square corners of the video and line pulse components.

The waveform resulting from a moving picture is considerably more

# VIDEO WAVEFORMS

complex, as shown by the example in Fig. 3.2. Here the scope is adjusted to give a line pulse in the screen centre with picture signal each side. Careful observation reveals some of the component parts of the picture signal. These look something like diffused sine waves. The 'ghost' waveform effect below black level results from the line pulse being locked to the scope X frequency while the picture components are drifting in and out of lock as their frequency changes with picture information. Slight rounding of the line sync pulse implies that the h.f. response was not as good as might be expected, but on the set the 2·5 MHz frequency bars on Test Card D were clearly defined.

While the angular shape of the waveform components may be present in the signal taken from the input to the video amplifier, distortion can be introduced by the video amplifier and will be shown on the video display picked up at say, the cathode of the picture tube. Such a display is shown in Fig. 3.3.

**Poor h.f. response**

This time the sync pulse is positive-going and the picture negative-going – because the cathode of a picture tube requires a negative picture drive to give, reciprocally, positive drive on the grid. The big feature of this waveform, however, is the very much rounded top of the line sync pulse. This shows a bad loss of h.f. caused by a considerable increase in value of the video amplifier anode load resistor. To give the full-screen vertical deflection the Y input was set to 10 V/cm (the waveform indicates a peak-to-peak signal of the order of 50 V) and the sweep to 10 μs/cm with expansion so as to detail the line pulse and individual components of the picture signal proper either side of the pulse.

The 'ghost' waveforms are also present on this display. Some are caused by the action of the retrace which, at high sweep velocities, may not always be completely suppressed by the blanking action of the scope. The retrace stroke, though far less intense than the forward stroke, then tends to produce a display of its own.

Sometimes the retrace is used purposely to present a display by switching off the blanking. In that way, a display of higher frequency signal can be observed, for the retrace is (or should be) much faster than the forward stroke; but the ratio reduces at high forward sweep velocities, as intimated above.

An interesting display is shown in Fig. 3.4. Here the line pulse is opened up even more by expansion. The reasonable shape of the corner on the pulse signal signifies improved h.f. performance. This is taken from the same set as Fig. 3.3, with the video amplifier anode load resistor restored to correct value.

The typical lack of picture definition resulting from a poor h.f. response (as in Fig. 3.3) appears in Fig. 3.5.

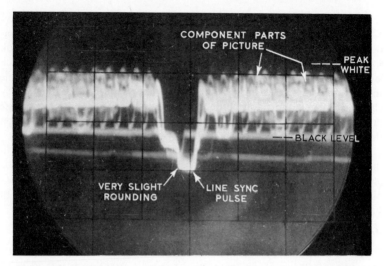

*Fig. 3.2.* Two half-lines of picture signal with line sync pulse between.

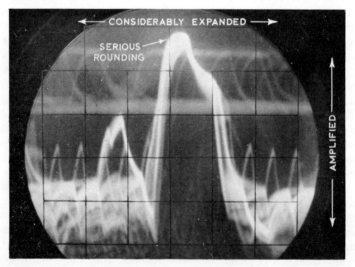

*Fig. 3.3.* Poor h.f. response revealed on video oscillogram.

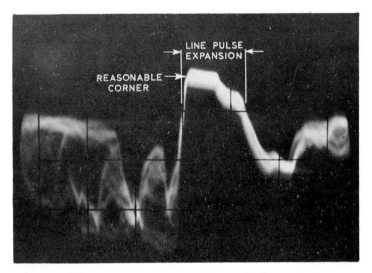

*Fig. 3.4.* Expanded line sync pulse, good h.f. response.

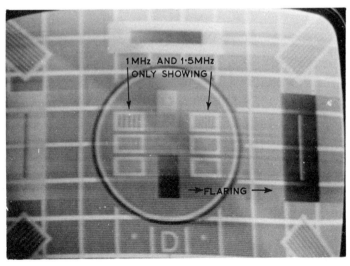

*Fig. 3.5.* Poor definition on Test Card D. Also notice flaring from blacks.

## SERVICING WITH THE OSCILLOSCOPE

Fig. 3.6 was taken from a set in which the line output transformer insulation had failed, resulting in severe corona discharges. The brightness control was turned back so that most of the picture disappeared, leaving only the very bright discharge effects.

*Fig. 3.6.* Severe corona effects due to a failing line output transformer.

**Corona interference**

From the defective set a video waveform display (Fig. 3.7) was picked up at the tube cathode. The sweep control was adjusted to produce four line sync pulses and the Y input set to 30 V/cm. The corona effect is clearly visible as very high amplitude pulses (about 120 V peak) rising from the sync pulses in the direction of picture white. Clearly, these 'interference' pulses occur at exactly the same time on each line scan. Since there are two (or more) pulses at each corona point, horizontally elongated, irregular white dashes are produced in a column usually towards the start of the scan, as the fault symptom picture shows.

What happens is that the corona discharges produce interference radiations which are picked up by the early stages (and sometimes the aerial, especially if a set-top type) and are thus present on the input signal. They are demodulated along with the video signal and appear at the tube.

Their high amplitude in the transformer sometimes results in radiation over relatively large distances and they can cause interference on a neighbour's set. If they happen to be picked up on a set receiving a different

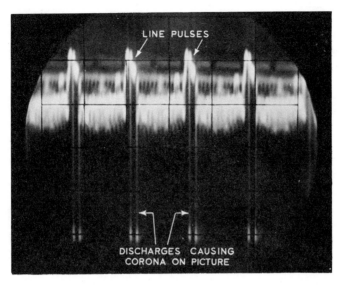

*Fig. 3.7.* Corona on video signal. Note this is 'synchronised' to the line sync pulses.

*Fig. 3.8.* Line timebase interference on the picture.

SERVICING WITH THE OSCILLOSCOPE

channel from that of the radiating set, the column of interference drifts at random across the screen of the affected set, producing the well-known 'windscreen wiper effect' (Fig. 3.8).

**Field sync period**

So far in this chapter, we have looked mainly at video waveforms displayed on a scope with the sweep set to trace out several complete lines of signal. By reducing the sweep velocity more and more lines of signal and sync pulses are displayed, until eventually all the line components merge into a strip 'chopped' by the field sync pulses. This happens when the sweep velocity is round 5 ms/cm.

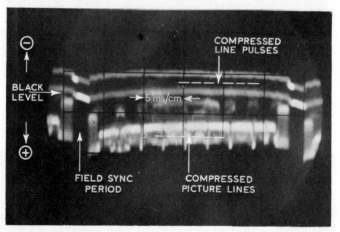

*Fig. 3.9.* Signal at tube cathode with sweep at field frequency.

Displays of field sync pulses can differ considerably from what may be expected. Line sync pulses have quite a resemblance to those sketched in textbooks, but the field pulses are less defined. Applying a sample of video signal to the Y input of a scope set to sweep at 5 ms/cm might bring a display similar to Fig. 3.9. This shows compressed line sync pulses at the top (positive-going) and compressed lines of picture signal (negative-going at the picture tube cathode) at the bottom. The field sync pulse is simply represented by a blank in the picture signal. Notice, however, that the line sync pulses continue during this period (as it is necessary to keep the line timebase synchronised).

Following a field, the signal goes to black level for the sync period. The 'blanks' in Fig. 3.9 constitute complete field sync periods which are com-

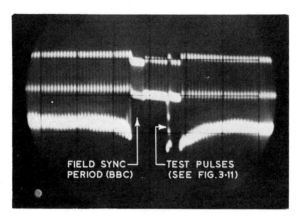

*Fig. 3.10.* Showing test pulses on a BBC transmission during the field interval (see Fig. 3.11).

*Fig. 3.11.* The pulses in Fig. 3.10 produce these signals between fields.

41

SERVICING WITH THE OSCILLOSCOPE

posed of a series of line pulses of black level datum. The waveform in Fig. 3.10 shows the field sync period in greater detail, and also a test pulse that is sometimes included during the field blanking period. This puts lines in the blank spaces between the picture frames, as shown in Fig. 3.11. The pulses are for test or control purposes only and add nothing to the actual picture information.

Of recent times more information is being encoded in these 'blank line' spaces for data transmissions that provide live-news/information services to those with suitable encoding equipment coupled to their receivers. The

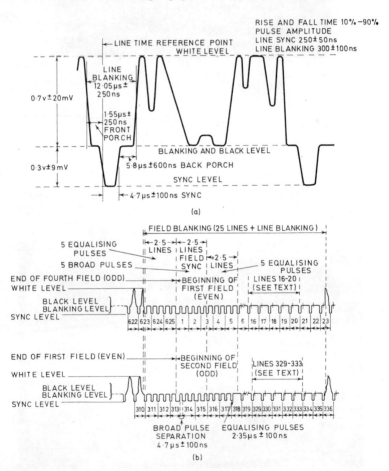

*Fig. 3.12.* Details of the IBA line and field 625-line signals at (*a*) and (*b*) respectively. See text for details.

42

pulses are digitally encoded on to the ordinary television signal, and both the BBC and IBA have developed such schemes, the former called *Ceefax* and the latter *Oracle*. (*Ceefax* information is transmitted at a rate of 7 megabits per second on lines 17 (330) and 18 (331) of the television waveform.)

Fig. 3.12 gives at (*a*) the waveform and parameters of a typical 625-line television signal. Pulse duration is measured at half-amplitude points. Blanking duration is also measured at half-amplitude points with a white level signal of line duration, which is the reason why the signal is shown starting and finishing at white level. At (*b*) is shown the vertical sync and blanking waveforms, with lines 7–14 and 320–327 omitted. Rise and fall times are measured between the 10 and 90 per cent amplitudes, and for field blanking are 300 $\pm$100 ns and for field sync and equalising $\pm$50 ns (1 ns = $10^{-9}$ s). The first and second fields are identical with the third and fourth fields in all respects except burst blanking on colour. Lines 16–20 may contain identification, control or test signals. Both (*a*) and (*b*) signals in Fig. 3.12 are from the IBA *Specification of Television Standards for 625-line System I Transmissions.*

When the picture signal falls to zero during the field sync period there is sometimes a minor change in black level due to the change in operation of the circuits, but with colour receivers a large change cannot be tolerated. Moreover, receivers with mean-level a.g.c. sometimes undergo a slight change in vision channel gain, though the relatively large time-constant of the a.g.c. feed circuits tends to minimise this. The overall effect of these changes may give the waveform a slight displacement either side of the field sync period, which is particularly noticeable in Fig. 3.9.

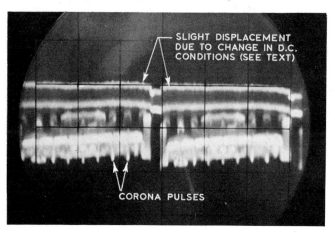

*Fig. 3.13.* Two field intervals, no expansion.

SERVICING WITH THE OSCILLOSCOPE

The effect is also shown in Fig. 3.13, which is the display of two sync periods (no X expansion). Note the pulses of interference on the picture signal at peak white on this waveform. They are from mild corona in the line output stage.

**Mains hum**

Mains hum can affect the video waveform when the scope is adjusted to suit the field repetition frequency (i.e. with the main 'sweep' control set to about 10 ms/cm). In Fig. 3.14 is shown a display of video signal on which there is rather a large 50 Hz hum voltage. The hum effect on the actual picture is shown in Fig. 3.15. The signal is locked on the screen by large-amplitude pulses during the field periods, not by the hum voltage. These pulses are derived from the field timebase generator due to back coupling to the point of waveform extraction.

At sweep velocities more suitable for the display of line signals and component parts of the vision signal slight hum does not interfere with the display unduly. Only a small fraction of the hum cycle occurs during the line trace and retrace period. However, excessive hum at high sweep velocities can cause a number of traces to appear on the screen one above the other, making it difficult to lock the display on one particular trace.

Vertical judder of a trace due to a small amount of hum is responsible for the thickening of the horizontal parts, relative to the vertical parts, of a waveform when photographed as explained in Chapter 1.

Difficulty is sometimes experienced in securing steady lock of a display owing to the scope failing to trigger at exactly the same part of the waveform on every trace. This happens particularly on changing video wave-

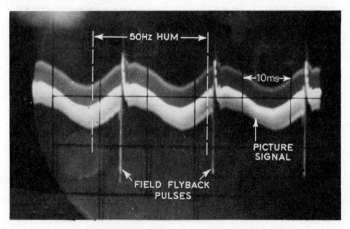

*Fig. 3.14.* Hum on video signal (also see Fig. 3.15).

## VIDEO WAVEFORMS

*Fig. 3.15.* The effect of hum on the television picture.

forms, the scope sync or triggering being affected by the varying amplitude of the picture signal components relative to the line sync pulse upon which the display may have originally been locked. Fig. 3.16 shows the resulting

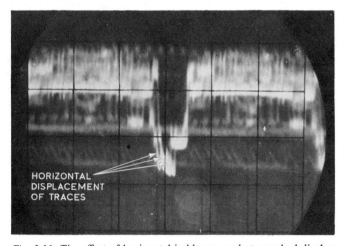

*Fig. 3.16.* The effect of horizontal judder on a photographed display.

45

horizontal displacement of traces. This oscillogram was taken with the film under exposure for several traces. Study of this picture will also reveal slight hum, causing the separate exposures of the line sync pulse, for instance, to be displaced not only horizontally, but also a little vertically.

### Video signal test points

The video signal, of course, can be picked up almost anywhere after the vision detector and, in this respect, the scope can be employed as a 'signal tracer'. Video signal amplitudes are sometimes given in makers' service manuals and on their circuits. These can be extremely useful in ascertaining the location of a video defect. Lack of contrast, for instance, could be caused by a host of faults, but the number of possibilities can be greatly reduced by checking the video signal peak-to-peak amplitude at various points.

Output of a vision detector is 2 to 3 V peak-to-peak (higher on some models) and the p–p signal at the cathode of the picture tube may be as high as 60 V. A considerably lower amplitude signal at the picture tube cathode (causing low contrast) could be due to lack of video drive at the video output valve grid (resulting from low detector signal) or a low emission video amplifier valve. A check of signal amplitudes with the scope against the service manual or circuit would quickly bring to light the area at fault.

Low signal at the detector output can well mean that the detector diode itself is faulty, especially if the sound is up to standard (indicating normal input signal) and if the gain of the vision i.f. channel appears fairly normal. Plenty of drive at the control grid of the video amplifier yet a lack of output at the anode and a correspondingly small signal at the picture tube cathode is a sure indication that something is amiss in the video amplifier stage proper.

Scope testing in video circuits is especially useful in sets featuring cathode-follower driver stages to the picture tube. In early receivers, the goodness of the standard-change switch can be quickly assessed (assuming inputs at both standards) without the labour of tracing the two sections wire by wire and component by component.

A great deal of fault-finding time can be saved by employing the scope for point-to-point signal location on printed circuit boards with the set operating in its cabinet. The video signal can be traced up to the picture tube and up to the sync separator.

Lack of video or weak signal at the input to the sync separator is a cause of impaired line and field synchronising. With a scope the origin of signal attenuation can be pinpointed very quickly and without the bother of removing components and disconnecting wires.

Hum on the video signal can be 'seen'. As this may be due to a heater/

# VIDEO WAVEFORMS

cathode fault in a video valve, or even in the sync separator, it may not be heard in the speaker. Hum in the video output valve (or even on the output, due to a picture tube defect) can affect the picture in a number of ways, but when hum finds its way into a line timebase of the flywheel type, a picture symptom like Fig. 3.17 is commonplace. It does not take long with the scope to find the point of hum injection – and its cause.

*Fig. 3.17.* The effect of hum in flywheel-controlled line timebase.

The scope can also be used to find video signal across decoupling capacitors when they are not working properly. These capacitors may be in the screen and cathode circuits of the video amplifier or driver valves – but remember that the cathode bypass capacitor may be applying compensation and not acting as a complete video-signal bypass.

The scope can also be used to check the goodness of decoupling on the vision a.g.c. line. There should be very little – if any – video signal here, but some of the lower frequency video components may be present on mean-level a.g.c. systems.

Just how much video is present at the output of the sync separator can also be revealed conclusively by the scope.

# 4: SYNCHRONISING WAVEFORMS

THE scope is particularly valuable for the analysis of waveforms in sync separator circuits and associated networks. When picture synchronising fails, for instance, we require to know what is (or is not) happening to the signals at the input and output of the sync separator stage, at the output of the field interlace filter (if fitted) or field integrator, and at the output of the line sync feed differentiator. A few tests with the scope will show conclusively the situation at these points and eliminate guesswork.

Scope testing around this area in more up-to-date receivers avoids damage to printed circuit boards and sub-assemblies. Too many disconnections, reconnections and component substitution tests, often demanded when sync fault diagnosis is attempted with a multimeter alone, are not good for printed circuit boards and, of course, are time-consuming.

It is desirable to work with a circuit diagram of the receiver and to have small test clips on the ends of the scope leads. All that is necessary is to establish on the circuit diagram the component and to which end of it the 'live' lead should be connected, and to clip this to the actual component wire on the circuit board – the part itself is usually identified on the board nowadays.

### Test points

In many cases, the 'earthy' scope lead should be connected to the metal chassis of the set (or to the h.t. negative line), leaving the 'live' lead for tests around the board. It is a good idea to terminate a second 'live' scope lead with a slim test prod of a type that has low capacitance to the hand. This facilitates speedy signal tracing around the set and allows connection to points not large enough to accommodate a clip. Sometimes it pays to solder small wire loops on the board component terminations to act as 'test points' to which the 'live' scope lead can be clipped. For detailed waveform observation a fixed connection is required; but much of the old trouble of locking a display on the screen is avoided if the scope has 'trigger sync' facilities.

Always have in mind that the chassis or h.t. negative of the set is probably connected direct to the mains supply. It is essential that this chassis connection be to mains *neutral*, for connecting the 'earthy' terminal of any instrument to live mains is simply asking for trouble.

# SYNCHRONISING WAVEFORMS

Many service workshops now have isolated mains feed to each bench point from fairly large double-wound transformers. This is the safest arrangement but it can give rise to hum troubles. Mains field from a transformer can put quite a ripple on the trace of a scope working close by. Much of this hum can be deleted by earthing the chassis (or h.t. negative). Indeed, this is often essential. However, care must be taken to ensure that a neighbouring operator using the same supply has not decided to earth his equipment on the opposite terminal!

In all cases the earthy scope connection must be connected *direct* to the chassis or h.t. negative of the equipment under test, not to any 'earth' on the set. If this is not done, ripple voltage developed across the capacitor used for d.c. isolation between chassis and earth socket makes waveform display difficult.

It is permissible to employ an isolating capacitor in series with the 'live' scope lead. Whether this lead can be screened depends upon the effect that the capacitance of a screened lead would have on the signals and circuit to which it is connected. The policy here is never to employ screened wire where it is possible to secure a reasonable ripple-free trace with a short length of unscreened wire. When high Y gains are necessary, the pick-up of residual signal on the input lead may be fairly high, and the presence of line scan signal radiated from the output section may demand screening of the scope input.

The circuit of a transistor sync separator as used in the Decca 25 series of colour receivers is given in Fig. 4.1(*a*). Video signal from an emitter follower stage is communicated to TR1 base through C1/C2/R1. The signal is negative-going on picture, which means that the n–p–n TR2, which is biased by R2/R3, conducts only on the positive-going sync pulses. At the collector, therefore, the sync pulses are negative-going. The input and output signals, along with their amplitude and sweep values, are also given on the diagram.

The line sync pulses are applied to a flywheel-controlled line oscillator through C3, which attenuates (due to its relatively low value) the field sync information, while the field sync pulses are applied to an integrating (adding) circuit through R4/C4/D1 etc., and thence to the grid of the first triode of a multivibrator field generator pair.

The composite line and field sync pulses are shown by the second output oscillogram of 2 ms/div sweep. The nature of the line sync pulses after passing through C3 to the flywheel discriminator diodes is also shown.

In the field feed circuit, the diode (D1) is biased to remove all but the field pulses, and the resulting field sync to the multivibrator is also shown on the diagram.

This is a relatively common sort of circuit nowadays, and one particular fault which might prove baffling is essentially 'solid' line lock but very weak field lock, affected by changes of picture content. This is almost always

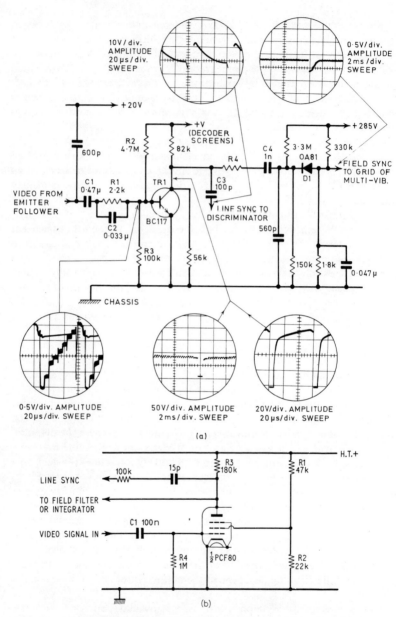

*Fig. 4.1.* (*a*) Circuit of the sync separator and line and field sync feeds used in the Decca 25 series of colour receivers. (*b*) An early type of valve sync separator stage.

## SYNCHRONISING WAVEFORMS

caused by open-circuit or increase in value of the 4·7 M resistor R2 biasing the sync separator transistor. A lot of time can be spent in analysing D1 and associated circuit when such a fault is present (as this would appear logically to be the part of the circuit in trouble), but to no avail.

Fig. 4.1(*b*) is the circuit of a typical valve sync separator stage. This uses the pentode section of a PCF80 triode-pentode valve. Video signal from the anode of the video amplifier valve is applied to the control grid, and the valve is biased towards anode current cut-off by the signal, causing charging of C1. The valve is given a short grid base by the use of a low screen grid potential (obtained from divider R1/R2) and by a relatively low anode potential.

With negative-going picture signal (positive-going sync pulses) anode current flows only during the sync pulses, so that the voltage developed

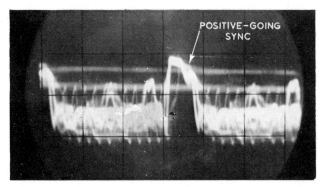

*Fig. 4.2.* Video signal at the input of sync separator stage with sweep for line.

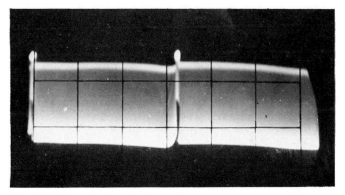

*Fig. 4.3.* Video signal at the input of sync separator stage with sweep for field.

across load resistor R3 takes the form of negative-going sync pulses. Oscillograms in Figs. 4.2 and 4.3 show the video signal at the control grid. Fig. 4.2, with a sweep of about 20 μs/div, shows the line pulse, while Fig. 4.3 with a slower sweep shows the field pulse period. These traces should be the first to be looked at when investigating sync troubles. If there is no video or low signal level at the sync input, then trouble exists somewhere in the video feed coupling.

Next thing to look at are the pulses at the output of the sync separator stage. If the sweep is adjusted to around 2 ms/div, the formation of the field sync pulse will be displayed, and X expansion can be used to get more detail, as shown in Fig. 4.4. This oscillogram shows almost complete absence of picture information, meaning that the sync separator is doing its job admirably.

By keeping the scope connections the same but increasing the sweep to about 20 μs/div, a line sync pulse can be resolved as in Fig. 4.5. The rounded corner implies h.f. attenuation, and this could be in the video channel, the sync separator stage or even in the scope (particularly if high signal shunt capacitance is present, which is why a low capacitance probe should be used). Poor h.f. response can be responsible for line pulling (to the right) on the television picture itself. The thick horizontal parts of the trace (Fig. 4.5) are caused in part by the reduced velocity of the spot and to the presence of a trace of picture information, but insufficient to impair the sync operation. The less bright inverted 'ghost' traces result from the flyback.

**Interpreting the oscillograms**

The sweep time indicated by the horizontal divisions on the graticule may not exactly correlate to the sweep time indicated by the sweep range switch. This is because the sweep is affected by the setting of the 'fine' sweep control, and one usually adjusts this either to synchronise the trace or to obtain the display of several complete signal cycles. To read the sweep in terms of time per graticule division (μs or ms/div or μs or ms/cm) it is necessary to set the 'fine' sweep control to the 'calibrate' end of its range, which is usually fully clockwise. Moreover, even in the 'calibrate' position of the 'fine' control the displayed time will be in error when the trace is expanded horizontally by the X expand control; thus, to read time from the graticule this control also needs to be set to 'calibrate', which is usually fully anticlockwise.

Similar reasoning applies to the volts per graticule division (V or mV/div or V or mV/cm). There is usually a 'fine' attenuator for the Y input as well as a calibrated switched attenuator, so for getting the correct voltage from the graticule the 'fine' attenuator needs to be set to 'calibrate', which is usually fully clockwise. Correction, of course, needs to be applied when the

# SYNCHRONISING WAVEFORMS

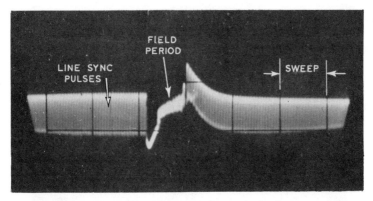

*Fig. 4.4.* Formation of the field sync pulse at anode of sync separator valve. This was obtained with X expansion. Note absence of picture signal.

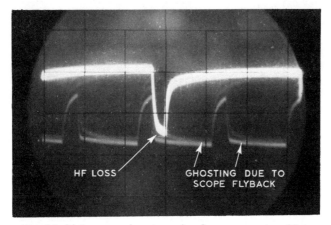

*Fig. 4.5.* Line sync pulse at anode of sync separator. Note the rounding of the corner, meaning h.f. loss. This could be responsible for pulling to right of picture on whites.

times 10 or times 100 button or switch is operated; also when a low-capacitance, voltage-divider probe is used at the X input (see Chapter 1).

## Z modulation

An alternative method of reading sweep time from a trace is by applying timing pulses from an accurate generator to the first grid of the c.r.t. This is called *Z modulation*. When a positive-going pulse is applied to the Z modulation input (usually at the rear of the scope) the trace will intensify for the

53

period of the pulse. Conversely, if the pulse is negative-going the trace will be darkened or blanked for the period of the pulse. In this way markers can be superimposed upon the trace in terms of beam modulation.

The oscillogram Fig. 4.6 was produced by X expansion with the scope connected as for Fig. 4.5. Here, however, the h.f. loss is considerably less, and this is the sort of line sync pulse display that one would expect to obtain from normal sync separator (and video circuit) operation. The rather thick trace is due to an abnormally high exposure time.

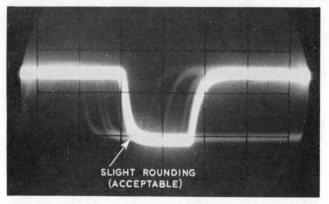

*Fig. 4.6.* Expanded line sync pulse at sync separator valve anode.

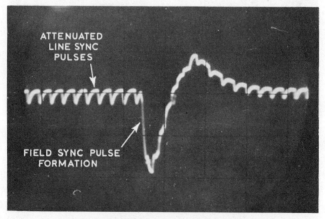

*Fig. 4.7.* Expanded field sync pulse formation after integrator. Note the reduced amplitude line pulses here.

## SYNCHRONISING WAVEFORMS

**Interlace hazard**

One would, of course, expect both field and line sync pulses to be present at the anode of the sync separator valve, as revealed in Fig. 4.4. However when extracting the field pulse signal from somewhere in the integrating section of the field sync circuit the line pulse amplitude should be considerably attenuated if the circuits are working properly. (One of the biggest hazards to correct interlace is the presence of line pulses in the vicinity of the field timebase generator.) But when checking sync pulses towards the point of application to their respective timebases, pulses generated in the timebases can confuse the issue. The best thing to do is to 'kill' the generators, as explained in Chapter 1. Fig. 4.7 shows a field sync pulse that stands well out from the line pulses.

**Interlace filter**

Fig. 4.8 shows the circuit of a typical interlace filter, this being connected between the anode of the sync generator and the sync input to the field timebase generator. In effect it is a low-pass filter that lets through only the field pulses (owing to their relatively low repetition frequency) while attenuating the rapid line sync pulses.

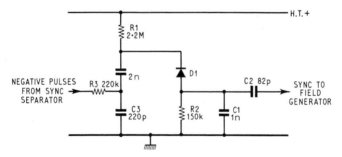

*Fig. 4.8.* Circuit of interlace filter using diode and charging circuit.

Diode D1 is biased to cut-off by resistors R1 and R2 across the h.t. supply. During the field sync pulses the diode is switched on by the negative-going pulses on the 'cathode'. Current flows in R2 and charges C1. As the time-constant of C1, R2 is large with respect to the intervals between the field pulses, a single, clean-cut field pulse is developed across the time-constant and is fed to the field generator through C2. Such a pulse is shown in Fig. 4.9 and on the horizontal part of this trace there is almost complete freedom from line pulses.

The waveform in Fig. 4.7 was obtained prior to the diode circuit but even here the network is sufficient to push down the line pulses. Note that R3 and C3 of Fig. 4.8 form an integrating circuit, which can be considered as a low-pass filter in itself. The great advantage of the diode, however, is clearly appreciated by comparing Fig. 4.7 with Fig. 4.9.

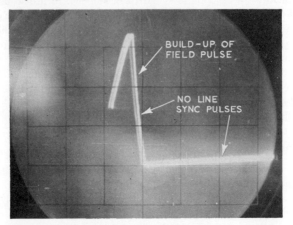

*Fig. 4.9.* Field sync pulse after interlace filter circuit.

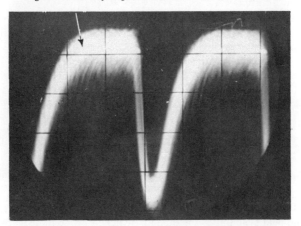

*Fig. 4.10.* Noise signal on field sync pulse (indicated by the arrow), after integrator and interlace filter.

The effect of noise on a field sync pulse is interesting to see. A filtered field sync pulse from a very weak BBC1 transmission is shown in Fig. 4.10. This shows considerable noise (often referred to as 'grass') on the pulse, implying that the sync separator is not receiving sufficient signal to make it limit properly. This is the cause of erratic line and field lock.

# SYNCHRONISING WAVEFORMS

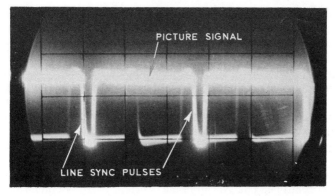

*Fig. 4.11.* Line sync pulses at anode of sync separator showing traces of picture signal.

*Fig. 4.12.* Pulling on picture lines as the signal content changes.

To recapitulate, the sync separator, receiving correct video signal, should delete all the picture signal; the network connecting the sync separator to the line timebase generator, being a high-pass filter (called a differentiator) should not pass any field sync to the line timebase; the network connecting

## SERVICING WITH THE OSCILLOSCOPE

the sync separator to the field timebase generator, an effective low-pass filter, should greatly attenuate line sync pulses and pass only the field pulses, adding these together to form one big pulse (Fig. 4.7).

Alterations in value or characteristics of the associated components can impair the performance. Failure of the interlace filter diode, for instance, can greatly reduce the amplitude of the field pulse and upset the vertical hold. A faulty integrating component can let through line pulses and impair the interlace performance. Trouble with a differentiating component can impair the horizontal lock. If the sync separator stage lets through picture signal, as shown by Fig. 4.11, the line lock performance is often affected, the lines 'pulling' as the picture signal content changes, as shown in Fig. 4.12.

Colour receivers, most dual-standard and some 405-line-only models use flywheel controlled line synchronising, where a phase discriminator compares the line sync pulses with sample line timebase signal. Phase deviations between the two signals yield a controlling potential for the line oscillator. However, let us begin a study of timebase waveforms with a 'recap' of the principles of direct sync. This uses just a sync separator stage and simple $RC$ filters between this and the timebase waveform generators.

### Sync networks

Fig. 4.13 represents the basic differentiator filter with a rectangular sync input impulse and a spiky output waveform. The degree to which the output differs from the input depends largely on the time-constant of the filter components $C$ and $R$.

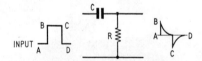

*Fig. 4.13.* Basic differentiator filter circuit with its rectangular sync pulse input and spiky output waveform.

Time-constant is a figure expressing the rate at which a capacitor charges, or discharges, through a resistor. As shown by Fig. 4.14, charging is rapid at first but slows down as the charge builds up in the capacitor, leaving a smaller voltage drop across the resistor and, hence, a smaller charging current. A discharge curve is similar but inverted. Multiplying capacitance in farads by resistance in ohms gives the time in seconds taken for charge or discharge to reach 63 per cent of completion. This is the 'time-constant' time. Virtually complete charging or discharging takes about $5CR$.

How does an output waveform as in Fig. 4.13 come about? When the input pulse arrives there is no charge in the capacitor and therefore no opposition to charging current except $R$. The almost instantaneous rise of voltage represented by the steep front of the pulse appears unchanged at the output. Now, while the input voltage remains constant, *C* charges, the voltage rises across it and the output voltage across *R* falls. Comes the sudden collapse of the input pulse and, as to all abrupt changes, *C* acts like a short circuit. The positive-to-zero input swing appears as a negative-going voltage at the output. A slower discharge gives the tail of the spike.

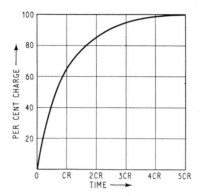

*Fig. 4.14.* Capacitor charging (or discharging) curve.

When *C* and *R* are changed over, as in Fig. 4.15, the filter becomes an 'integrator'. The initial short-circuiting action of *C* kills the sharp front of the input pulse. The output may rise and fall just as a succession of charge and discharge 'cycles'.

As already suggested, the precise effect of a differentiator (or integrator) circuit depends on the relationship of pulse frequency and duration to the filter's time-constant. Obviously, what happens to a signal applied to the filter is affected by the charge or discharge, if any, left over from previous

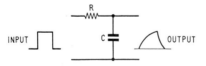

*Fig. 4.15.* Basic integrator used in the field sync feed circuit.

events. If the time-constant is long compared to pulse duration, output of the type in Fig. 4.16(*a*) is obtained from a differentiation; (*b*) is the kind of result obtained with a short time-constant.

The differentiator is designed to give sharp spikes to trigger the line timebase but between sync separator and field timebase an integrator filter is

## SERVICING WITH THE OSCILLOSCOPE

used. As this smooths out brief pulses, the line sync pulses are eliminated but the long field pulses have effect. In fact, the circuit 'integrates' the train of pulses (see Fig. 4.17) because the pulses are longer than the intervals.

The field timebase triggers when the voltage across $C$ reaches a predetermined level, usually established by a diode circuit. The performance of such a filter is shown by oscillogram Fig. 4.9.

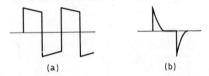

*Fig. 4.16.* Output of integrator (*a*) of long time-constant and (*b*) of short time-constant compared with the pulse repetition frequency.

Field sync pulses are also applied to the differentiator and their leading and trailing edges produce 'spikes' that keep the line timebase in sync during field flyback; these pulses are responsible for proper line interlace.

Although, actually, it is the generator oscillations which vary with respect to the accurately-timed sync pulses, the out-of-sync state of affairs is as if the sync pulses were running up or down the timebase waveform (Fig. 4.18). When a pulse arrives towards the top of the waveform the oscillation is triggered. Providing the generator is adjusted to run very nearly at sync frequency, the sync pulses can give the 'flip' each time round.

*Fig. 4.17.* Integrated series of field sync pulses, forming one large pulse.

This explains why, on the field sync for instance, a slightly slow or fast generator causes the picture to roll in a series of jerks, locking for a short time while the sync pulses arrive around the top of the waveform. A roll without jerks indicates lack of, or severe attenuation of, sync pulses.

What happens when the sync pulse arrives at the top of the waveform depends to some extent on the particular circuit but often, as indicated in Fig. 4.18, the grid or base rises above cut-off, starting conduction in the valve or transistor and instigating the flyback at the end of a scan.

# SYNCHRONISING WAVEFORMS

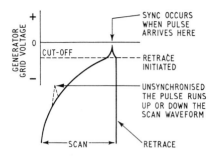

*Fig. 4.18.* Showing how sync pulses trigger the timebase generator (see text).

This is direct sync, which is nearly always employed in the field circuit. The brief, frequent line sync pulses, however, are more easily distorted and misrepresented by interference and random noise pulses. To avoid the ragged sort of scanning this causes, the flywheel form of line sync is widely adopted. On the 405-line standard the sync separator tends to suppress pulses but, with the 'inverted' waveform of the 625-line signal, interference pulses are in the same direction as the sync pulses and the flywheel circuit becomes all the more desirable.

Some 625-line sets feature a special 'interference cancelling' sync separator which deletes the spurious pulses. This is effective field-wise but less so from the line aspect.

**Flywheel sync**

The indirect or flywheel method of synchronising employs a discriminator circuit which receives the shaped line sync pulses together with a sample of the pulses produced by the line generator or output stage. When these two sets of pulses are in step the output from the discriminator is zero (or nearly so). Should the generator run a little fast or slow the two sets of pulses fall out of step and a positive or negative voltage is developed at the output of the discriminator.

This voltage is employed to slow down or speed up the line generator as required to bring it into step with the sync pulses. The voltage may control the line generator directly, by effectively adjusting its time-constant (i.e. the grid voltage), or it may vary the value of an 'electronic reactance' which, being across the line oscillator frequency-determining circuit, changes the frequency a little as required.

A circuit of relatively long time-constant (this is the 'flywheel') is interposed between the discriminator output and the line oscillator (or electronic reactance) so that, should the line sync pulses become distorted or fail

## SERVICING WITH THE OSCILLOSCOPE

completely for a line or two, the generator remains at the correct speed. Some systems feature an *LC* circuit tuned to the line repetition frequency (switched over the two line standards) which helps maintain accurately-timed scanning.

The ragged vertical effect is eliminated and the worst that happens is a slight waving of vertical content under extreme interference.

Such a circuit, employing a pair of rectifiers in the discriminator, is given in Fig. 4.19. The rectifiers are fed from the secondary of a transformer the primary of which is in the sync separator anode circuit. Sample line pulses from the line output stage, shaped by C2 and R1, are fed to the centre of the diode pair via C1. Circuits using transistors or integrated circuits are similar in operation.

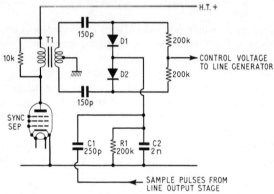

*Fig. 4.19.* Showing how the control voltage is obtained for a line flywheel-controlled timebase oscillator.

The scope is useful for checking the balance of the discriminator circuit relative to chassis, and the oscillogram Fig. 4.20(*a*) shows the waveforms at one side of the secondary winding of T1 (Fig. 4.19) while Fig. 4.20(*b*) shows the waveform at the other side.

The Y input was set to 10 V/div and the sweep was adjusted to display two pulses. Thus it can be seen that the pulse peak-to-peak amplitude is about 30 V, fairly well balanced at each side. A short- or open-circuit (or weakness) in one rectifier would show up as unbalance of the two waveforms. Unbalance in the discriminator produces poor line locking and a tendency for the line to flick out of lock on changes of camera or picture content.

The pulses shown are mainly derived from the line output stage via C1. The line sync pulses without the line timebase pulses are shown in Fig. 4.21. Note the complete lack of picture signal, the fairly broad horizontal trace being caused not by picture signal but by film over-exposure.

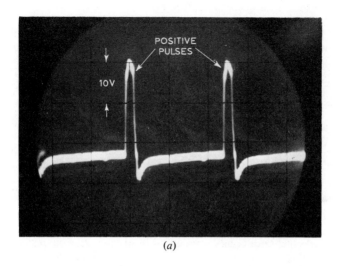

(a)

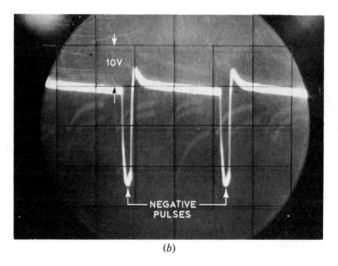

(b)

*Fig. 4.20.* Oscillograms of waveforms each side of the flywheel line discriminator, (*a*) on one side and (*b*) on the other. Correct balance of these two is essential for proper operation of the circuit.

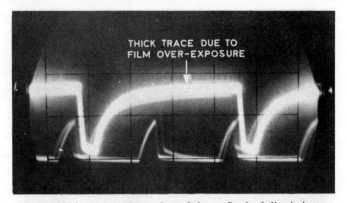

*Fig. 4.21.* Line sync pulses only as fed to a flywheel discriminator.

Note the complete lack of picture signal, the fairly broad horizontal trace being caused not by the picture signal but by film over-exposure.

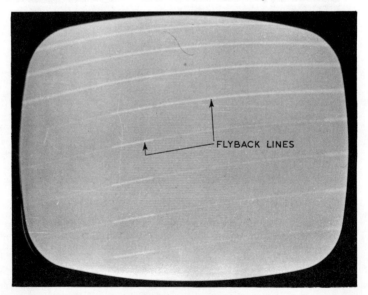

*Fig. 4.22.* Display of flyback lines on synchronized raster due to failure of blanking circuit.

## Blanking waveforms

Modern sets employ a network for blanking the field flyback lines from the raster or picture. Early sets without this feature or modern sets in which something is wrong with the network show the flyback lines on the screen, see Fig. 4.22. Theoretically, the tube beam should be cut-off by the video signal falling to black-level during the field sync period. In practice, however, this rarely happens due to loss of the d.c. component and the need to set the tube bias a little above black-level (by turning up the brightness control) for a picture of balanced contrast.

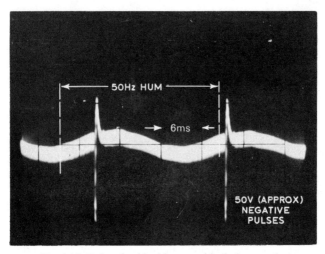

*Fig. 4.23.* Pulses for blanking at grid of picture tube.

The blanking network operates by feeding the negative-going pulses which occur in the field timebase during the retrace periods back to the tube grid. Thus during the periods when the flyback lines would normally be visible the large negative pulses push the tube hard into beam current cut-off. Blanking pulses of this nature, as taken from the control grid of the picture tube, are shown in Fig. 4.23. These can be up to 50 V peak amplitude in the negative direction. The wave on this display is caused by the presence of 50 Hz hum, while the thickness of the wave results from the pick-up of pulses from the line output stage. This, in fact, is how the waveform may appear *in practice*, revealing beyond doubt the presence of field blanking pulses. The timebase in the scope is running at about 6 ms/div sweep, with the fine sweep control set for a stationary display.

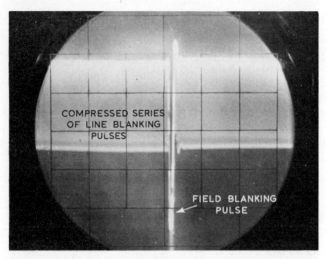

*Fig. 4.24.* Line pulses and a field blanking pulse at picture tube grid.

Generally line pulses also are fed back to the tube grid to provide line flyback blanking. The display Fig. 4.24 (sweep 1 ms/cm and Y-input 10 V/cm) shows these as a steady band of illumination, the field pulse dropping out below the negative-going line pulses.

When blanking pulses are applied to the tube grid, the grid is loaded with a resistor of about 1 M or less between it and the slider of the brightness control. This resistor is necessary to avoid the pulses being short-circuited by the brightness control. The blanking pulses are applied to the grid via *RC* coupling.

# 5: TIMEBASE WAVEFORMS

WHAT we may call the end product of the field timebase is a current, in the field scanning coils, having a waveform or shape suitable for deflecting the scanning spot linearly from top to bottom of the screen and then returning it swiftly to the top for the commencement of another scan.

**Field timebase**

An oscilloscope displays voltage waveforms and to depict a current waveform the current must be passed through a non-reactive load (i.e. pure resistance) to develop a voltage that can be monitored on the scope. The resistor must be of sufficiently low value to prevent affecting the normal operation of the field output stage, but not so low that the voltage cannot be shown on the scope without excessive Y-amplifier gain. High gain here might result in mains hum interference on the trace.

A 5- or 10-ohm resistor in the 'earthy' side of the field scan coil circuit generally satisfies these conditions. The waveform shown in Fig. 5.1 was obtained this way. It shows a reasonably linear scanning stroke, with just a little curvature at the start and a small reduction of retrace velocity towards the end of each flyback.

Textbook sawtooth waveforms are rarely obtained, mainly because some scan stroke non-linearity is required due to the geometry of the tube face and scanning angle. This is called 'S' correction.

Fig. 5.2 shows a waveform when vertical linearity controls were adjusted to give a picture with an expanded top as in Fig. 5.3. The steeper slopes of the waveform correspond to the parts of the picture where the lines are more widely spaced. Distortion and non-linearity of this nature can be caused by trouble in the negative feedback circuits in the field output stage. Increase in value of a resistor or an open-circuit capacitor are typical causes.

If the retrace is delayed too much, the picture information may start a fraction of a second before the scanning stroke. Some tolerance is provided during the field sync period by a number of final black lines, giving a broad, black horizontal band between pictures. Since this period is sometimes occupied by test pulses (see waveform Fig. 3.10), these will show as short horizontal lines and a dot at the top of the picture if the retrace slows down too much towards its termination. The test pulses or the Ceefax or Oracle

SERVICING WITH THE OSCILLOSCOPE

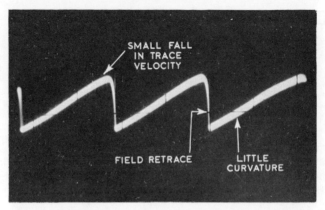

*Fig. 5.1.* Field timebase waveform as obtained from across a low-value resistor in the scan coil circuit.

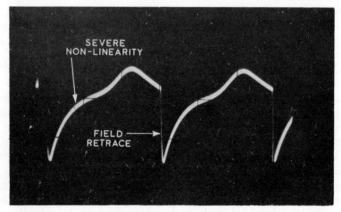

*Fig. 5.2.* Field waveform non-linearity due to mis-setting of the vertical linearity presets.

digital pulses (see Chapter 3) may appear in 'duplicate' at the top of the picture since two fields comprise one complete, interlaced picture.

The *voltage* waveform at the anode of the field output valve can be displayed by connecting the Y input of the scope to this electrode via an isolating capacitor (with the 'earthy' side of scope connected to chassis). The waveform here, though, tells less about linearity conditions than the scanning coil current waveform for, when a sawtooth current is driven through an induct-

# TIMEBASE WAVEFORMS

*Fig. 5.3.* Severe vertical distortion of the top of the picture due to the scan waveform at Fig. 5.2.

ance, the voltage itself is not purely sawtooth. The waveform at the output valve anode is shown in Fig. 5.4. The voltage pulse rises sharply at the start of the scan stroke and then holds almost flat during the scanning period, to fall rapidly on the retrace.

The rapid change of current in the inductive scanning elements during the retrace gives rise to an induced voltage pulse similar to that due to the line retrace. This can be clearly seen rising upwards on the waveform (Fig. 5.4). Here the Y input is set to 30 V/cm, so the average peak-to-peak amplitude of the waveform proper is about 60 V, but the induced voltage pulse goes off the screen at the top – it can rise to about 1,000 V in some receivers.

This is one of the reasons why field output valves have to have high peak-volt anode insulation and why, as the insulation at the anode tag of the field output valveholder deteriorates with heat and dust, there is a tendency towards flash-over, even though the d.c. at the anode may only be about 200 V. Suppression of this pulse is sometimes achieved by connection of a voltage-dependent resistor across the primary of the field output transformer. This sort of resistor has a high value when the voltage across it is little more than the peak-to-peak value of the nominal waveform, but its value falls considerably when the voltage rises during flyback.

Fig. 5.5 shows a field output valve anode waveform on a 100 V/cm Y

SERVICING WITH THE OSCILLOSCOPE

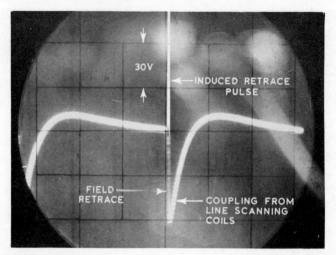

*Fig. 5.4.* Showing field retrace pulse.

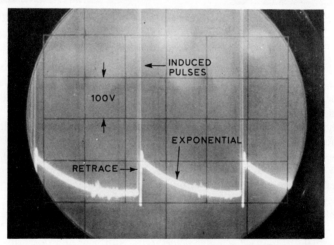

*Fig. 5.5.* Anode waveform of field output valve, showing induced pulses.

scale. This shows the flyback pulses rising well above 500 V peak-to-peak amplitude. Note also that on this waveform the scanning stroke is more exponential than square-wave. The exact nature of the scanning stroke of the trace is governed by the type of circuit and the setting of the vertical linearity control(s).

## Line-to-field coupling

The field waveform sometimes shows signs of signal coupled in from the line scanning coils. The little dots, for instance, on the high velocity part of the scanning stroke in Fig. 5.4 are due to this. They look almost like timing pulses. The retrace is also slightly affected. The effect on the television screen due to this is shown in Fig. 5.6.

## Output stage cathode waveforms

In some field output stages the waveform at the cathode can give clues as to whether the stage is working as it should. A little negative feedback may be featured in the form of an unbypassed resistor in the cathode circuit. One valve-type receiver, for example, has a 33 ohm unbypassed resistor in series with a 500 ohm resistor across which is connected the usual high-value electrolytic decoupling capacitor (500 µF in this case). The 33 ohm resistor improves the linearity and the waveform across the cathode circuit as a whole is shown in Fig. 5.7. The peak-to-peak amplitude of the waveform is 3 V, the Y input being set to 1 V/cm. Note the good linearity of this waveform, almost as good as the current waveform in Fig. 5.1, but with a small tendency towards an opposite curvature.

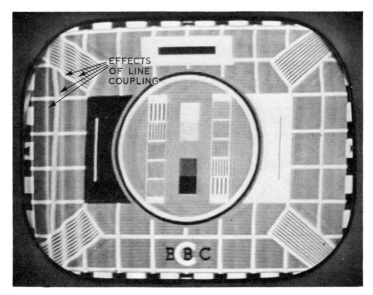

*Fig. 5.6.* The effects of line signal coupling into the field scanning coils on a picture.

## SERVICING WITH THE OSCILLOSCOPE

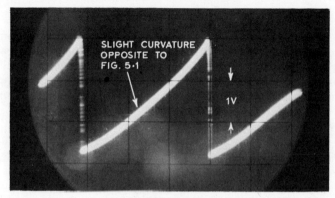

*Fig. 5.7.* Field voltage across unbypassed section of cathode circuit. In some sets this is parabolic (see Fig. 5.8).

The waveform shown in Fig. 5.8 was obtained from the cathode circuit of a field output stage from which a degree of feedback voltage is taken to the control grid circuit via the linearity control network. This is virtually a linearising waveform which, of course, has no direct resemblance to the sawtooth waveform expected in field output stages. This sort of signal is used for dynamic convergence in colour receivers, and is given the required tilt by the addition of sawtooth waveform (see Fig. 8.12).

### Field generator waveforms

The drive waveform from a field multivibrator is shown in Fig. 5.9. This is produced by the charging of a capacitor through the anode load resistor of the second stage of the multivibrator, giving the exponential charging stroke,

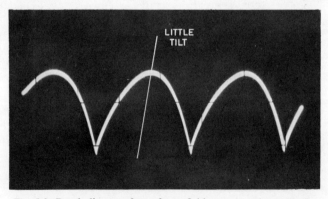

*Fig. 5.8.* Parabolic waveform from field output valve cathode. The tilt is due to the presence of some sawtooth voltage.

# TIMEBASE WAVEFORMS

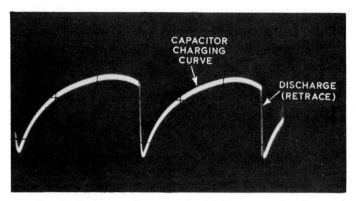

*Fig. 5.9.* Drive waveform at grid of field output valve. The exponential shape is due to the charging of a capacitor.

and the sudden discharge caused by the switching-on of the other stage. It will be seen that the charging waveform diminishes a little from its peak value: this results from the feedback linearising waveform, of which the waveform in Fig. 5.8 is a component.

The waveform at the grid of a field generator is shown in Fig. 5.10. The top of the waveform corresponds approximately to valve cut-off bias, and at values below this, of course, the valve is well into the cut-off region. The exponential rise from negative towards cut-off results from the discharging of the grid capacitor during the time the oscillator is cut off.

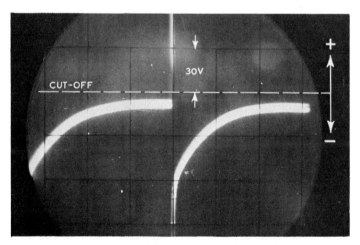

*Fig. 5.10.* Waveform at grid of multivibrator.

73

## SERVICING WITH THE OSCILLOSCOPE

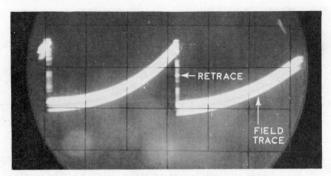

*Fig. 5.11.* Field drive waveform (also see Fig. 5.9).

When the capacitor loses sufficient charge to switch the valve on, the inductive feedback element, in the case of a blocking oscillator with anode-to-grid coupling transformer, causes the stage vigorously to swing through a cycle of oscillation, the effect of which immediately charges the grid capacitor again and holds the valve at cut-off during the discharge time of the grid capacitor.

In effect, the valve switches on only for a brief period of the scanning cycle. It is this action 'reflected in' the anode circuit that usually discharges the timebase charging capacitor which itself is charged through a resistor. The voltage across the capacitor of this second charging circuit represents the field drive waveform, the retrace occurring when the valve switches on to discharge the capacitor.

The drive waveform may appear similar to that in Fig. 5.9 or it may look more like Fig. 5.11, depending on the nature of a corrective waveform, if present.

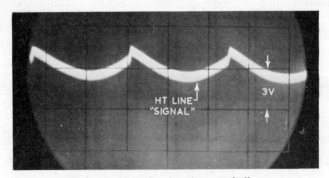

*Fig. 5.12.* Field signal on h.t. supply line.

The Y input is scaled at 30 V/cm in Fig. 5.10, giving a peak-to-peak (charge-to-discharge) amplitude of around 60 V. This waveform also shows positive and negative voltage pulses due to induced e.m.f. (during the retrace) across the grid winding of a blocking oscillator transformer. The drive waveform at the output of a multivibrator or blocking oscillator may be about 90 V peak-to-peak but is less than this at the control grid of the field output valve owing to the action of the coupling network and correction waveforms.

Timebase generators thus generally feature two charging circuits, one which fixes the repetition frequency of the generator (whose time-constant is adjustable by the hold control) and the other producing the drive waveform for the amplifier. Trouble in the former affects the speed of the timebase and prevents correct locking; lack of field amplitude and poor linearity are typical of timebase charging circuit trouble.

**Field buzz**

While the line frequency is readily attenuated in the sound circuits and common h.t. feed by ordinary filter and decoupling techniques, the lower-frequency field signal is not as easily suppressed. Traces of field buzz (as distinct from vision-interference-on-sound) can often be heard in the background when the sound modulation is low.

This can cause complaints when it rises a little above the acceptable level of about $-40$ dB, and it is then necessary to trace the point of entry into the sound channel. The 50 Hz signal usually gets into the audio stages proper either by common impedance coupling or by failure or value reduction of an electrolytic capacitor on the h.t. feed to the sound audio section or field timebase circuit.

A 50 Hz sine wave signal is less of a problem than the actual timebase signal, which contains many harmonics of the fundamental field frequency. Moreover, as we have seen, quite a heavy disturbance occurs in the circuits during the field flyback.

In spite of the large-value electrolytic capacitors on the h.t. feeding the timebase, it is surprising just how much field signal can be present. Fig. 5.12 shows the nature of the field signal on the poorly bypassed h.t. feed to a field amplifier. The Y input is scaled to 3 V/cm, thereby revealing a peak-to-peak field signal of about 3 V. It was found possible to reduce this by about 3 dB by improving the smoothing and decoupling.

**Line timebase**

The most exacting section of the receiver is the line timebase. Peak performance is demanded to deflect the electron beam fully across wide scan angles and to generate high-amplitude peak voltages which, after rectification, deliver an e.h.t. supply of 16 kV or more for the picture tube.

The higher e.h.t. of modern sets 'stiffens', so to speak, the beam and calls for greater scanning power. Dual-standard design requires peak performance to be maintained at both 10,125 Hz for 405 lines and 15,625 Hz for 625 lines.

Efficiency deterioration in the line timebase soon shows as reduced line scan, possible non-linearity and poor e.h.t. regulation on one or both standards. A limited amount of information concerning the performance of the line timebase can be gleaned from meter testing but almost nothing regarding pulse performance. An oscilloscope is essential to determine what is really happening from the signal aspect.

**Line drive**

First, a test can be made to prove that the line timebase generator is sending drive signal to the amplifier. Because the drive signal is of fairly high amplitude (or should be) a substantial amount of Y attenuation is needed. The signal should be coupled to the Y input through the isolating capacitor in the instrument or through an external 0·1 µF capacitor. A screened connecting lead is hardly necessary as the line signal is far stronger than any spurious hum pick-up.

Waveform Fig. 5.13 shows a typical 405-line drive signal. Vertical deflection (Y input) is scaled to 100 V/cm and the X base to 10 µs/cm, with the 'fine sweep' control adjusted for the locked display of two cycles. Drive amplitude at the control grid of the amplifier is about 150 V p–p, which is a fairly typical value, but it could be a little more or a little less without fault being indicated.

Fig. 5.14 shows the waveform at the same point but this time with the line timebase switched to 625 lines. The waveshape is similar, but the amplitude is a little greater, being about 160 V p–p. It does not always follow, however, that the drive amplitude is less on 405 lines. The converse could be true, depending on the exact nature of the circuits.

The drive *amplitude* rather than the wave shape (considered later) is the prime feature of this test. And the value of the test is greatly enhanced if the amplitude for the set is known. This is sometimes given on the circuit or in the service manual.

A component which often comes under suspicion in the line timebase is the line output transformer and its conclusive test often demands replacement with a component known to be in good condition. This is time-consuming and costly and the problem can sometimes be solved by checking the drive amplitude at the control grid of the line output valve. At least this test proves whether the circuits up to the amplifier valve are working correctly (also see page 91).

For example, if the drive amplitude is substantially below that given in the specification or known reference, then the trouble is likely to exist in the generator and/or coupling rather than in the output valve anode circuit. It

# TIMEBASE WAVEFORMS

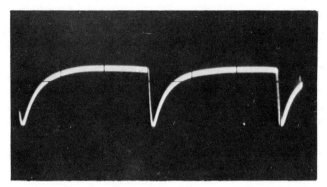

*Fig. 5.13.* Drive waveform at line output valve's signal grid on 405 lines. Y = 100 V/cm and X = 10 μs/cm.

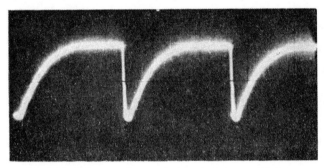

*Fig. 5.14.* Drive waveform as in Fig. 5.13 but at 625 lines.

should be noted, though, that the drive amplitude at the amplifier valve control grid can be influenced to some extent by the anode loading (i.e. by the line output transformer) and by the condition of the amplifier valve itself. The waveforms in Figs. 5.13 and 5.14 were obtained with the output stage in good order and with the line amplifier 'absorbing' the drive signal in the correct manner.

If the drive amplitude is correct within about ±5 per cent, yet the line output stage fails to work correctly (i.e. lack of e.h.t.), tests should be made of the d.c. voltage on the output valve screen grid (not at the anode direct). If this is normal and the h.t. supply from the booster diode circuit to the primary of the line output transformer is correct and the primary to the anode of the line output valve has continuity, then shorting turns in the line output transformer or in an inductive component coupled to it (like the width or linearity inductor or scanning coils) could be causing the trouble.

# SERVICING WITH THE OSCILLOSCOPE

**Output stage waveforms**

It is possible to obtain a waveform direct at the anode of the line output valve but as the peak voltage here rises extremely high (3–4 kV) instrument damage can result unless precautions are taken to attenuate the signal before feeding it to the Y input. Such attenuation is not difficult but it is often a problem to determine the exact value of attenuation applied by practical 'service bench' methods so that sense can be made of the measured amplitudes of the displays.

Large amplitude displays of line signal can be obtained simply by clipping the Y-input lead to the insulation of the cable from the transformer to the anode of the line output valve. The kind of waveform so obtained is seen in Fig. 5.15. This shows that during the scanning stroke the voltage is relatively constant. The heavy peaks occur during the retrace (flyback) when the current in the line amplifier inductive elements changes very suddenly.

Two such pulses (positive-going) are shown in Fig. 5.15. These rise to about 120 V peak on the oscilloscope (Y input scaled to 30 V/cm) but, of course, their real amplitude is much in excess of this, having in mind the considerable attenuation offered to the signal by the nature of its coupling to the Y input (that is, through the capacitance of the line output valve anode lead insulation!).

The oscillogram also shows that the pulses rise immediately from the base level of the waveform at the commencement of the retrace, yet fail to return immediately to the same level after the retrace. What happens is that the inductive elements of the anode circuit oscillate or 'ring' at their resonant frequency. This causes the ripple effect at the start of the scanning stroke part of the waveform.

'Rings' of this kind are responsible for the alternate dark and light vertical bars sometimes visible on the left-hand side of the television picture (Fig. 5.16). Not all sets exhibit the effect to any alarming degree though some early models gave quite a bit of trouble. The lines result from the changing velocity of the beam during its trace. Early sets suppressed the 'ringing' by special damping circuits. Modern sets are arranged so that this energy released during the flyback is added to the scanning stroke.

The retrace pulse switches off the boost diode and lets through to the line scanning coils the initial 'ringing' energy. This is responsible for the commencing deflection of the electron beam on the scanning stoke. After this energy has been exhausted, the diode switches on, damps any further 'ringing' and effectively couples the line output valve to the scanning coils, via the line output transformer, the valve contributing the remainder of the scan stroke.

Reversal of current in the line output transformer thus occurs about halfway over the scan stroke. Moreover, the action of the boost diode, and the

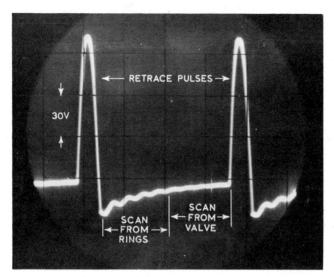

*Fig. 5.15.* Line pulse picked up by clipping the Y lead to valve anode cable insulation.

*Fig. 5.16.* Showing the effect of line timebase rings on a picture.

way it is arranged in the circuit, cause a capacitor (the boost reservoir capacitor) to charge up. This charge is also a product of the ringing action and its voltage is added to the ordinary h.t. line voltage, thereby giving boosted h.t. of up to 700 V or more to supply the first anode of the picture tube and, sometimes, the field timebase generator.

Low-loss line output transformer cores and tuned windings encourage 'controlled' ringing and contribute to the efficiency of the timebase. An extra aid towards improved efficiency is a technique of tuning the line output transformer so that the ring frequency is almost three times the retrace repetition frequency. This action causes the retrace pulses at the line output valve anode to be more of the nature of a 'trough' than a peak while encouraging those at the anode of the e.h.t. rectifier (coupled from the line output transformer overwind) to rise to even greater amplitude. The net result is that the e.h.t. performance is enhanced while the anode of the line amplifier valve is subjected to reduced peak voltage.

### Screen grid waveform

While it is not possible to measure the pulse amplitude directly (or very accurately) at the anode of the line amplifier valve, it is possible to glean a fair idea of how the amplifier proper is working – and responding to the line drive – by looking at the waveform at the screen grid. Fig. 5.17 shows a waveform at this point.

This cannot be considered as typical, as the wave shape is influenced by the nature of the amplifier circuit and feedback on the screen grid circuit. However, it will be seen that it has much in common with the anode waveform of Fig. 5.15. The main difference is its amplitude, Fig. 5.17 being scaled to 10 V/cm on the Y input, the signal being at 405 lines. Thus here the p–p amplitude is about 30 V.

Another waveform at this point on a different set working at 625 lines is shown in Fig. 5.18. This is scaled at 30 V/cm, thereby revealing an overall amplitude of about 140 V. Another waveform is shown in Fig. 5.19, again scaled to 30 V/cm, giving an amplitude of about 120 V p–p.

### Shape of drive waveform

At this juncture we should look back to the drive waveforms in Figs. 5.13 and 5.14 in terms of shape (earlier we discussed only the drive waveform amplitude). The classic textbook drive waveform shape is the sawtooth, so it may come as rather a surprise to see the severe exponential shape of the waveforms mentioned.

The waveforms in Figs. 5.13 and 5.14 were derived from the output of the line generator via the coupling to the line amplifier. In some sets the generator output waveforms are exactly the same as the amplifier drive waveforms, while other models employ pulse shaping networks in the

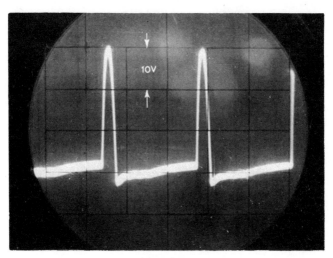

*Fig. 5.17.* Line waveform at screen grid of line output valve.

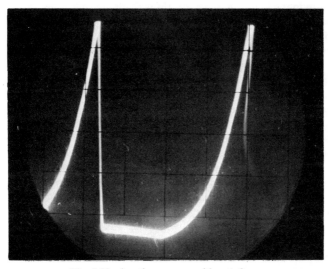

*Fig. 5.18.* Another screen-grid waveform.

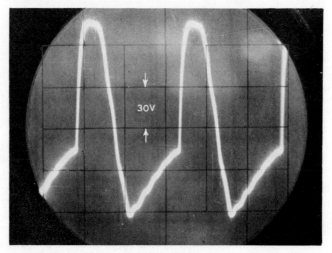

*Fig. 5.19.* Another screen-grid waveform. The nature of the waveform at this point is governed by the design of the line output stage.

coupling from the generator to the amplifier. Nevertheless, the drive waveform in the line circuits does not differ greatly from those depicted.

It must be remembered that the requirement is for a sawtooth current waveform in the line scanning coils. If the scanning coils were wholly resistive the demanded voltage waveforms across them would also be sawtooth. For purely inductive coils, the voltage waveform would be rectangular (something like a square wave with marks and spaces to match the scanning and retrace periods respectively).

In practice line coils are made up predominantly of inductance plus resistance, with a little shunt capacitance thrown in. This means that the voltage needed across the coils to drive a sawtooth current through them is a compromise between rectangular and sawtooth. This anode-circuit condition is 'reflected' in the waveforms in Figs. 5.13 and 5.14, having in mind also that the output stage itself incorporates linearity correcting devices which to some extent modify the drive voltage requirements.

Looking again at these waveforms, during the initial period the scanning current is being supplied by ring energy. When this energy is exhausted the line output valve is made fully conductive by the drive waveform at that time starting its stable, maximum amplitude trace. This is rather the same as the top of a rectangular wave. Thus, current continues to change in the line scanning coils in a linear manner, giving the remainder of the sawtooth current supplied via the output valve itself.

# TIMEBASE WAVEFORMS

At the end of the scanning stroke the drive abruptly falls negative, cutting off the line amplifier valve off, causing the rapid change of current in the inductive elements of the circuit and the retrace pulse described earlier.

**Line generators**

Line generators are generally either of blocking-oscillator or multi-vibrator type. Each sort can differ considerably in detail from one set to another however. Instead of the conventional multivibrator using a couple of triodes, a circuit may use a triode and a pentode, the latter sometimes being the line output valve. A blocking oscillator may employ a triode or sometimes a pentode driving the line output valve. In some sets the line output valve may also be arranged as oscillator (self-oscillating line output stage).

Dual-standard sets sometimes adopt yet different arrangements. With flywheel-controlled line sync a sine-wave oscillator may be used to drive a shaping circuit prior to the line output valve. This was a favourite of several makers a decade back, and has appeared again in some of the 625-line monochrome and colour receivers. Valves are still used but the trend is towards the use of transistors and in some cases integrated circuits. However, the principles involved can be appreciated from the valve circuits, and these are being retained in part in this second edition, including dual-standard circuits.

Another dual-standard technique is the use of a blocking oscillator with the time-constant (line hold) circuit fed from the anode of a separate triode valve, this valve picking up its control grid signal from a flywheel line sync

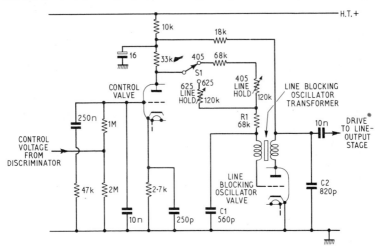

*Fig. 5.20.* Line blocking oscillator type of generator controlled flywheel-wise from discriminating voltage.

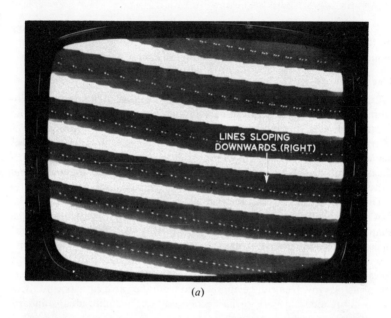

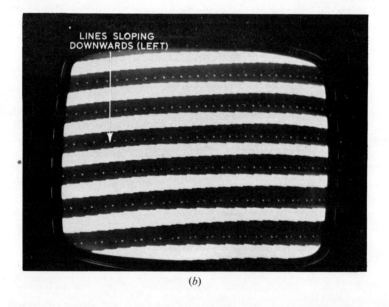

*Fig. 5.21.* Line generator running too fast (*a*) and too slow (*b*).

discriminator circuit. When the sampled line pulses from the output stage coincide with the sync pulses the discriminator gives zero (or a nominal) output. However, when the oscillator drifts relative to the sync pulses, a plus or minus output is produced and it is this which is fed to the grid of the valve mentioned above. The valve's conductivity is altered and a change in anode voltage results. This change is reflected in the line blocking oscillator stage, bringing the line pulses back into step with the sync pulses.

A circuit of this kind is shown in Fig. 5.20. Standard change switch S1 selects the appropriate line hold control. These controls are 'buffered' so that the correct locking point falls towards the centre of rotation. The oscillator frequency is determined by the time-constant C1, R1 and the resistance of the selected hold control circuit in conjunction with the potential applied at the anode of the control valve. As this potential is altered, so the frequency of the oscillator is changed and, as we have already seen, the potential is under the control of the voltage applied to the control grid from the flywheel sync discriminator.

**Line generator faults**

Typical troubles in this type of oscillator (as, indeed, in other types) are drifting of the locking point and inability to lock within the range of the appropriate hold control. If the hold control is hard against one of its stops and the lock is still not quite there, the nature of the line break-up can give a clue as to the whereabouts of the trouble.

If the picture appears to drift horizontally across the screen, not quite locking, when the hold control approaches the normal locking point, the discriminator and control valve circuits should be studied. However, if the effect is like (*a*) or (*b*) in Fig. 5.21, the generator proper is running too fast or too slow.

The too-fast effect at (*a*) shows the break-up lines sloping a little downwards to the right, while the slope is opposite at (*b*) when the timebase is a little slow. Effect (*a*) indicates a decrease in the value of a time-constant resistor or capacitor, while (*b*) indicates an increase in value of either or both.

Relative to Fig. 5.20, both effects can also be caused by a change in control valve conductivity, resulting from a change in valve emission or in the value of an associated resistor, or even from trouble in the discriminator or its feed circuit to the control grid.

Fig. 5.22 shows the symptom when the line oscillator is running at half its correct speed. Here two almost complete pictures are displayed side by side. Severe line compression is, of course, in evidence and non-linearity is exaggerated, there being in Fig. 5.22 more compression on the right-hand picture than on the left-hand one. This is to some extent due to the incorrect line speed, for it will be recalled that the line oscillator serves essentially to discharge a capacitor to give the retrace, while its charging through a resistor

## SERVICING WITH THE OSCILLOSCOPE

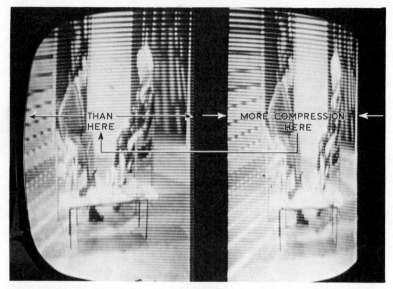

*Fig. 5.22.* Line running at half speed. Note uneven compression at right and left.

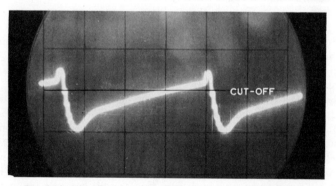

*Fig. 5.23.* Waveform at grid of line blocking oscillator valve.

gives the forward stroke, this effect resulting in the line drive waveform. Thus there is some relationship between the time-constant of the charging circuit and the line speed and this is destroyed when the line locks at half speed.

Fig. 5.22 symptom is not very common unless a resistor increases in value by almost half its correct value or if the time-constant oscillator capacitor does likewise. Either of these troubles, of course, makes the time-constant

# TIMEBASE WAVEFORMS

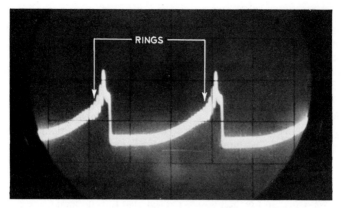

*Fig. 5.24.* Voltage across C1 (Fig. 5.20) on 405 lines.

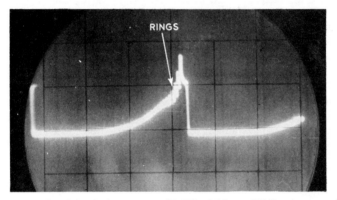

*Fig. 5.25.* Voltage across C1 (Fig. 5.20) on 625 lines.

longer and thus reduces the oscillator frequency. The symptom can, however, sometimes happen on a dual-standard set if the oscillator switch goes wrong and the frequency fails to rise from 10,125 Hz to 15,625 Hz when changing from 405 to 625 lines.

Fig. 5.23 is the waveform at the grid of a line blocking oscillator. The top of the rising waveform represents the grid cut-off potential, while the pips at the top rise to zero grid voltage. The lower part of the waveform is well in the negative grid region.

Looking at Fig. 5.20, let us suppose that current is rising through the anode winding of the blocking oscillator transformer. It rises at a rate controlled by the inductance of the winding. An e.m.f. is induced in the grid winding, rising from a negative towards a positive value, as shown by waveform Fig. 5.23. This continues until either the valve or the transformer saturates. Due to the diode action of grid and cathode, grid current flows and charges C1.

The effect is that the control grid rapidly becomes negative with respect to cathode, causing the valve to block by going well into cut-off. The capacitor discharges through R1 and associated line hold control resistances until conduction of the valve again takes place, repeating the cycle.

The waveforms shown in Figs. 5.24 and 5.25 show the voltage across C1 on 405 lines and 625 lines respectively. Notice the small amount of ringing at the start of the charging cycle. This is caused by self-oscillation in the blocking oscillator transformer proper. Fortunately, this has very little effect on the drive waveform – which is derived from the charging of a capacitor (C2 in Fig. 5.20) in the anode circuit of the oscillator.

The waveform in Fig. 5.26 shows the signal at the anode of a line blocking oscillator, which is based on the action of the charging capacitor from which the drive signal is fed to the line output valve. This is the sort of waveform obtained during a servicing operation and it differs somewhat from the classic anode waveform of a blocking oscillator devoid of coupling and charging components.

This waveform shows how the charge builds up exponentially across the charging capacitor (Fig. 5.20, C2) and how the charge is very rapidly exhausted by the sudden conduction of the blocking oscillator. Without the charging capacitor connected, the anode waveform would be similar, but without the exponential charging cycle. The swing from a negative to a positive value would be almost vertical.

In some circuits the anode waveform may appear rather like the capacitor waveforms shown in Figs. 5.24 and 5.25.

Fig. 5.27 shows the waveform at the anode of a multivibrator. The rise in value is, again, very much exponential because of the charging capacitor. In the multivibrator case, though, the capacitor usually charges from the h.t. line through the anode load resistor of the second stage. The waveform shows that the capacitor charges fully before the valve conducts due to the oscillatory action, conduction then being indicated by the swift fall from a positive to a negative value, giving the line retrace or flyback. The pip at the top of the waveform is a reflection at the anode due to the sudden build-up of voltage at the valve's grid, giving the conduction pulse.

**Self-oscillating line timebase**

Fig. 5.28 gives the circuit of a self-oscillating type of line output stage. Actually, the output valve (V2) here is not totally self-oscillating because feedback is provided by way of triode section V1. This is a cross between a multivibrator and blocking oscillator circuit.

The anode of V1 is coupled to the control grid of V2 via C1, while the

# TIMEBASE WAVEFORMS

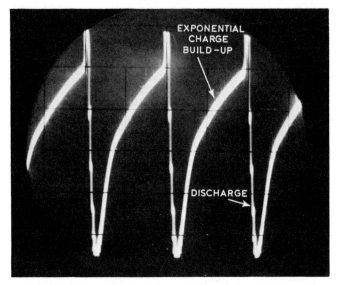

*Fig. 5.26.* Signal at anode of line blocking oscillator valve.

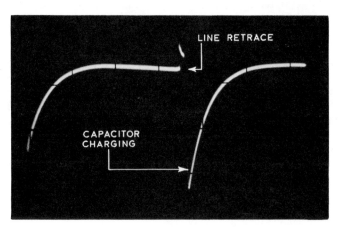

*Fig. 5.27.* Signal at anode of line multivibrator valve.

control grid of V1 is coupled to the screen grid of V2 through C2. Oscillations are sustained by the feedback from the line output transformer also to the control grid of V1 via C3.

The drive waveform is derived essentially by C4 charging through C1 and R1. C4, being the smaller value, acquires a charge before C1. When V1 is

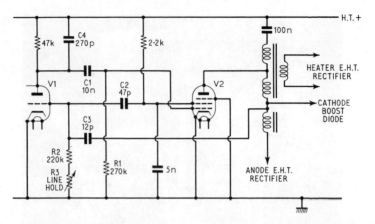

*Fig. 5.28.* Circuit of self-oscillating line output stage, using a triode in the feedback circuit. This is a cross between a multivibrator and a blocking oscillator.

suddenly switched on by the oscillatory action, C4 is rapidly discharged and this gives the retrace.

The oscillator repetition frequency is determined by the time-constant provided in the main by C3 and R2 and R3 in series. The frequency is adjustable, giving the line hold control, by R3 being variable.

The waveform at the anode of V1 in this circuit is similar to that shown in Fig. 5.27, but the precise wave shape depends on the exact nature of the circuit employed. The shape of line oscillator waveforms is not especially important in practical servicing, for mostly one wishes to know whether or not the stage is producing a drive signal. If it is not, then the question, why not? calls for some idea of how the circuit should function.

The oscillator may be producing a signal but of amplitude insufficient to work the output stage properly. In such a case, there may be barely sufficient e.h.t. to give a raster. Or a raster may be produced at very low setting on the brightness control, the e.h.t. collapsing completely when the beam current requirements are increased by advancing the brightness control. The raster (or picture) 'blows up', then extinguishes.

Drive amplitudes are often given in the servicing data or on the circuit diagram. A typical value is 150 V peak-to-peak at the oscillator anode. Knowing that the drive amplitude (and waveform, as far as is known) is correct, then lack of e.h.t. voltage could well point to trouble in the line output stage.

It is possible to employ the oscilloscope for shorted-turns checks on the line output transformer and such tests will now be considered.

## Line output transformer test

When a symptom indicates lack of e.h.t. the condition of the line output transformer is questioned. This often inaccessible component is connected in circuit by a host of wires and is rarely the easiest of parts to check by substitution. If it happens to be one of the plug-in transformers designed for swift change, the chances are that a replacement is not immediately to hand. Since it is reasoned that the transformer may not be the culprit, anyway, there is often some reluctance to put the set aside, pending arrival of the replacement part, without first making a few circuit tests.

Such tests in the line timebase are best carried out with an oscilloscope as already explained. Trouble in the line output transformer is responsible for a high proportion of line timebase and e.h.t. faults, insulation collapse being a common example. A short circuit between one or two or even a half-dozen adjacent turns causes no significant drop in the overall d.c. resistance of a winding and so a suspect transformer may show the specified winding resistances and yet be faulty. Insulation and d.c. resistance tests, of course, are useful to reveal the state of the insulation between the various windings and from windings to core.

The line output transformer is a highly-developed component which works in the way that it does because of its low losses. Any slight impairment in efficiency, therefore, will show up as a line or e.h.t. fault, and even a single shorting turn can have a dire effect on performance. This can be readily demonstrated by winding a turn of insulated wire round the transformer core and then shorting the ends together. In some sets this will sabotage the e.h.t. completely. In others it may simply reduce width, impair e.h.t. regulation or distort the line scan depending on the nature of the set and on the sophistication of the line circuit.

Shorting turns in the windings themselves mostly kill the line output stage operation completely. In some cases the line whistle will just be heard (this is handy because it indicates conclusively that the line generator is working, at least), at other times it may be increased in intensity, but in all cases the line output valve runs pretty hot, often abnormally so due to the impaired anode loading.

The oscilloscope can be employed to get an idea as to the relative goodness of a line output transformer. The idea is to get the transformer to 'ring' by applying pulses across one of its windings while watching the waveform on the oscilloscope.

Any winding on a transformer, choke or scanning coil has a natural frequency of oscillation – the resonant frequency. This is due to the inductance of the winding and the capacitance distributed across it, which can be considered as a lumped capacitance in parallel with the winding. If disturbed electrically, such a circuit will 'ring'; that is, oscillate at its natural frequency at progressively reducing amplitude. The amplitude of oscillation and the time taken for it to decay to zero are governed by the efficiency (or $Q$ factor). A high-$Q$ circuit will ring for a longer period than a low-$Q$ circuit or a damped high-$Q$ circuit. Amplitude of oscillation of a high-$Q$ circuit is generally greater than in a low-$Q$ one, other factors being equal.

The windings on a line output transformer are normally fairly high $Q$ (low loss) and are prone to ringing. Indeed, we have already seen how a ringing line output transformer, or associated inductor, can upset the operation of the line timebase. When such a winding is heavily damped by a short-circuited turn, the ringing is small to start with and dies rapidly.

Now, if we can get a winding on a suspect line output transformer to ring and can monitor this on an oscilloscope and compare the amplitude and general characteristics with a ring produced by a similar transformer known to be in good order, we will have an unambiguous test for 'goodness'.

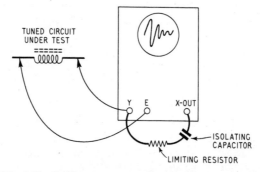

*Fig. 5.29.* Oscilloscope connections for obtaining rings.

This can be done (see Fig. 5.29) easily by connecting the winding between the earth and Y-input terminals of the scope and then subjecting the winding to a current or voltage pulse synchronised to the X sweep. With a high-$Q$ circuit input, sufficient disturbance is sometimes created simply by scratching the blade of a screwdriver on the Y terminal side of the circuit. If plenty of Y gain is applied, a damped oscillation will be seen to traverse the screen.

A more satisfactory way is to arrange for the oscilloscope's timebase to provide the pulses. At the end of the forward trace the X waveform falls rapidly during the retrace to take up position for the next trace. At that time a pulse is produced by the timebase and this, being synchronised to the X sweep, is ideal for producing rings.

Some oscilloscopes deliver sufficient pulse voltage at the *external sync* connecting terminal. It is then simply necessary to connect this circuit direct

# TIMEBASE WAVEFORMS

to the Y input, via a loading resistor and/or capacitor depending on the type of instrument employed, and set the Y gain and X sweep for the best results.

The instruments used by the author, however, deliver insufficient pulse voltage at the sync terminal, and it is found best to use the signal present at the X-output terminal. There is adequate pulse amplitude here to ring even the lowest of $Q$ circuits. It is only necessary to feed the signal to the Y side of the winding through a series resistor and capacitor. These components serve as isolation and also reduce the load across the ringing circuit. Tests can be made to establish the best values to use for any particular application, and the values used for testing line output transformers should certainly be noted and retained. The set-up described is shown in Fig. 5.29.

Investigations have shown that it is not necessary to disconnect the windings of the line output transformer from the receiver circuits to make ringing tests. One of the best places to connect the Y input is between the line output valve and the booster diode cathode; that is, across the top caps of the two valves in the screened line output section.

The oscillogram shown in Fig. 5.30 shows the ringing display obtained in the 405-line position and Fig. 5.31 shows how the display was slightly altered when the switch was changed to the 625-line position. Y input and X sweep were set to 3 V/cm and 30 µs/cm for both displays. Fig. 5.32 shows what happened to the display in Fig. 5.30 when both halves of the line scanning coils were shorted. The effect here is basically that of a reduction in

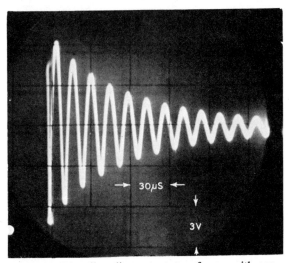

*Fig. 5.30.* Ring from line output transformer with connections to scope between line output valve anode and booster diode cathode. This is on the 405-line position.

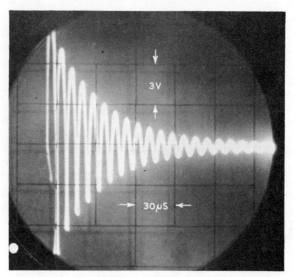

*Fig. 5.31.* Ring as in Fig. 5.30, but obtained on the 625-line position.

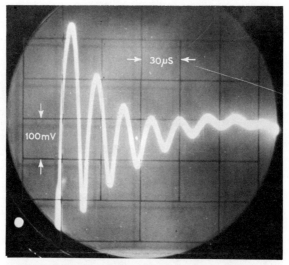

*Fig. 5.32.* This ring was obtained with the set-up as in Fig. 5.30 and when both halves of the line scanning coils were shorted. There is also a severe fall in ring amplitude.

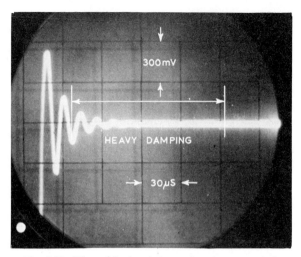

*Fig. 5.33.* Ring with shorting turn in e.h.t. overwind. Note the heavy damping.

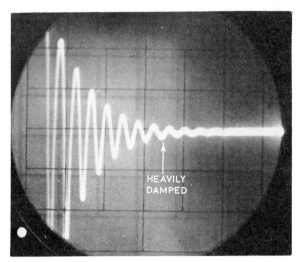

*Fig. 5.34.* Ring when two turns of wire were wound round core of line output transformer and shorted.

ringing frequency. There was also a severe fall in amplitude and, in fact, to retain a reasonable amplitude the Y input had to be transferred to across the anode of the e.h.t. rectifier and the cathode of the booster diode.

It really does not matter which winding is used to show the rings, for trouble on any one of the windings will reflect across any winding connected to the Y input. It has been discovered that the maximum ring amplitude is obtained from the winding with the maximum turns, such as the e.h.t. overwind, and that both this winding and the primary proper (between booster diode cathode and line output valve anode) are highly sensitive to transformer loading in terms of shorting turns.

Fig. 5.33 shows how the ring rapidly decays when a shorting turn exists in the e.h.t. overwind. The amplitude is also some ten times below that of the oscillograms in Figs. 5.30 and 5.31, extra Y gain being applied to obtain sufficient vertical deflection. Fig. 5.34 was obtained across the primary winding when two turns of wire were wound round the core and the ends shorted; and the waveform in Fig. 5.35 was produced when the primary was short circuited, the ring being taken from across the e.h.t. overwind.

These traces clearly show the value of an oscilloscope for ringing tests in the line output transformer. One could, in fact, be absolutely certain in transformer appraisal after making such tests had displays with a correctly

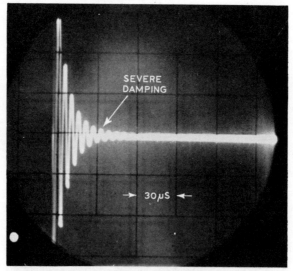

*Fig. 5.35.* Here the ring was taken from across the e.h.t. overwind and the transformer primary short-circuited. Note the slightly higher ring frequency and the very severe damping.

working transformer been obtained previously. In some instances, one could be fairly sure a transformer was suffering from shorting turns even without experience of a 'norm'.

It does not take much time to compile a small book of transformer ringing data on the various makes of set that pass through the workshop. A small sketch of the waveform should be included with notes on where the Y input was connected, the sweep and Y-input settings and the settings of the fine gain and sweep controls; also the value of the resistor and/or capacitor connected in series with the pulsing signal from the sync or X-output terminal.

A ringing test can be applied to any h.f. or a.f. choke or transformer and also to the scan coils. One or two tests using the oscilloscope will soon show its value.

### Testing in transistor circuits

An example of scope testing in a transistorised field timebase of the Thorn 9000 series colour chassis is given by the circuit in Fig. 5.36 and the oscillograms in Fig. 5.37. The oscillograms were obtained with a 10 M/11 pF 10:1 probe (see Fig. 1.8) and with the receiver set up for normal reception and display of a colour bar signal. The amplitude and time scales of the oscillograms are the actual values with the attenuation of the probe taken into account.

Sync signal is applied to FR1/1 and an integrator circuit is formed with this resistor and C401, the signal then going to TR402 base through W401–C402. In the absence of sync W401 is reverse-biased, which means that under this condition the field oscillator runs free.

The field oscillator comprises the complementary pair TR401/TR402, and during the scan both transistors are non-conducting owing to the large positive charge on C404, which causes reverse-biasing of W402 such that the $V_{be}$ of TR401 is zero, and since there is no forward bias on TR402 base this transistor also is non-conducting.

During this time C404 slowly discharges through FR1/4 until towards the end of the scan W402 and TR401 become slightly forward-biased. The resulting $I_c$ of TR401 then causes current to flow through FR1/6 into TR402 base, such that this transistor is turned on, which initiates the retrace. This action is regenerative and soon both transistors are fully bottomed. During the retrace C404 begins to charge through FR2/4, the emitter-base of TR401, W402 and FR2/6 until the base current of TR401 approaches zero. This causes TR401 $I_c$ to fall, thereby bringing TR402 out of bottoming. The resulting increase in TR402 collector voltage is coupled by C404 to W402, which causes this diode to be reverse-biased up to 10 V. The action is again regenerative until both transistors are fully off, when C404 begins to discharge through FR1/4 and the field scan once more commences. The action, of course, is timed by the integrated field sync information.

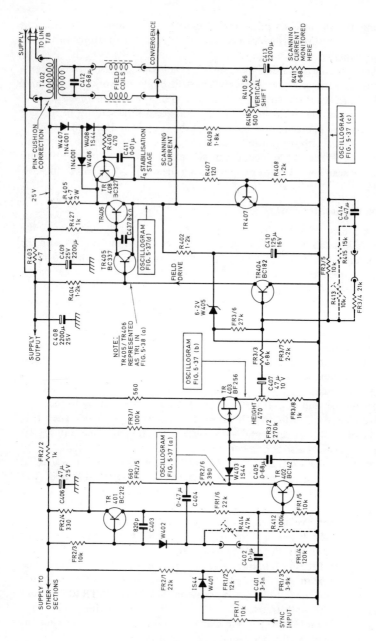

*Fig. 5.36.* Circuit diagram of field timebase section of Thorn (British Radio Corporation) 9000 series colour chassis. See text for description.

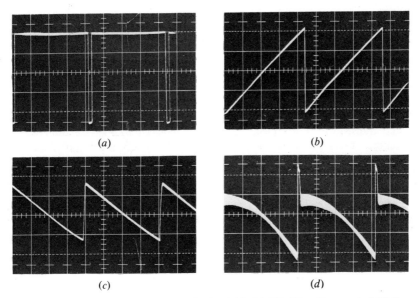

*Fig. 5.37.* Oscillograms in circuit of Fig. 5.36. (*a*) TR402 collector at 5 V/div amplitude, 5 ms/div sweep. (*b*) TR403 source at 1 V/div amplitude, 5 ms/div sweep. (*c*) Field current at 5 V/div amplitude, 5 ms/div sweep. (*d*) Field output signal voltage at 5 V/div amplitude, 5 ms/div sweep.

In other words the field oscillator works rather like a switch, such that the switch is off during the scanning stroke and on for a short period during retrace. Oscillogram Fig. 5.37(*a*) clearly shows the action at TR402 collector, where the amplitude is scaled to 5 V/div and the time to 5 ms/div.

The field sawtooth generator is operated by the switching waveform of Fig. 5.37(*a*), such that at the start of the scan C405 is discharged, but then begins to charge as the scan commences *via* FR3/1. This resistor and FR3/2 give a 13 V divider source of 68 k, and it is these components that determine the amplitude and linearity of the sawtooth waveform. Because the sawtooth amplitude is large compared with the source voltage, the non-linearity is excessive. The sawtooth is, as will be appreciated, exponential, such that as its amplitude increases so it deviates more and more from a linear ramp (sawtooth). This characteristic is employed to provide 'S' correction during the second half of the scan.

The sawtooth is fed to the field effect transistor (f.e.t.) TR403 arranged in source-follower mode. This gives a low output impedance, which is required for driving the linearity correction circuits and the field amplifier. The waveform at TR403 source is given at Fig. 5.37(*b*), which is scaled at 1 V/div amplitude and 5 ms/div sweep.

The scanning current is monitored by the 0·68-ohm resistor R411 (bottom right-hand corner of the diagram) and is fed back through FR3/5, FR3/4 and C414. These components introduce mild differentiation, and the waveform which is then subtracted from the output of TR403 yields the start of the 'S' correction. The waveform fed back to TR403 is shown at Fig. 5.37(c), which is scaled to 5 V/div vertical and 5 ms/div sweep (horizontal).

The output stage can be simplified to the circuit in Fig. 5.38(a), where all the components not concerned with the circuit operation have been deleted and transistors TR405/TR406 have been replaced by the one transistor denoted TR1. This shows how the quiescent current of the output stage is stabilised, since as the output voltage swing is very nearly equal to the h.t.

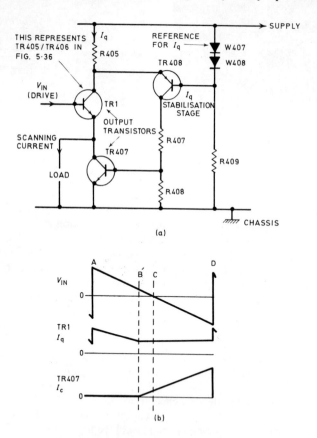

*Fig. 5.38.* (a) Simplified field output stage of Fig. 5.36, and (b) associated waveforms. Text gives description.

# TIMEBASE WAVEFORMS

it is important that the quiescent conditions are accurately maintained to avoid output limiting.

The two diodes W407/W408 are forward-biased by R409 to produce a reference voltage. The quiescent current flows through R405 and the voltage resulting across the resistor is monitored by TR408 and compared with the forward voltage drop of W407. If the quiescent current is low R405 voltage will also be low, which causes TR408 to pass more current until stability results when R405 voltage is equal to the forward voltage of W407 (W408 need not be considered in this discussion).

Sawtooth drive is applied to TR1 base, and from point A to B (Fig. 5.38(b)) the transistor is conducting and passing current through R405 and the load (i.e. field scanning coils). TR408 is reverse-biased, which means that both TR407 and TR408 are non-conducting. At point B the current has reduced to the quiescent current ($I_q$) and any further reduction will cause TR408 and TR407 to conduct. At point C the scanning current must fall to zero, but from B to D TR1 remains at $I_q$. This means that TR407 needs to conduct from point B so that at point C, where the scanning current is zero, both transistors run at $I_q$, the load current then being zero. From point B the load current reverses – flowing into TR407. As the input voltage falls TR1 endeavours to pass less current, which causes TR408/TR407 to turn on harder, thereby bringing about an increase in scanning current.

Clearly, the circuit contains a number of refinements which are outside the scope of this particular book to investigate in detail (see, for example, the author's *Newnes Colour TV Servicing Manuals*, Newnes–Butterworths).

In Fig. 5.36 the output voltage is taken via R402 and integrated by C410 to eliminate a.c. feedback. From the voltage here, 6·2 V is subtracted by the zener W405, and the resultant voltage is stabilised by the $V_{be}$ of TR404 as a reference.

The signal voltage at TR406/TR407 junction (i.e. across the load point shown in Fig. 5.38(a)) is given by oscillogram Fig. 5.37(d). Since this is a voltage waveform it is, of course, of exponential nature. This is scaled at 5 V/div amplitude and 5 ms/div sweep.

# 6: TELEVISION TESTS FOR HUM, DISTORTION AND RESPONSE

THE waveforms so far considered have been mostly those of signals produced by the receiver itself, and have included demodulated aerial signals, amplified signals of this kind, timebase signals and signals modified by shaping networks and so forth. As long as the set under test is able to respond to broadcast signals, this sort of oscilloscope testing is accurate and time-saving.

The receiver is 'disturbed' only by the connection of the scope. If this is carried out reasonably well, by the methods explained in earlier chapters, the fault condition should not change appreciably from that present without the instrument connected. This is not always the case when fault tracing with a testmeter, especially one that requires the test circuit to supply all the power to move the meter's pointer.

Previous chapters tell how normally-present signals can be displayed. It is possible also to use the scope to detect signals that should not be present. Testing for spurious hum signals in the h.t. circuits has already been considered but the idea can be extended to checking for hum in the sound, video or timebase circuits due to causes other than poor h.t. smoothing.

Impaired heater-cathode insulation in a valve, or the picture tube, is a common cause of hum trouble that fails to respond to additional h.t. line smoothing. Trial-and-error valve replacement is the customary check. However, a scope will quickly show where a hum signal is gaining admittance. It is only necessary to connect the Y-amplifier input to the cathode of each suspect valve in turn after first having set up the scope to display a low-level 50 Hz waveform. To test that the scope is correctly adjusted, the Y input can be connected to a valve heater tag, when a full-amplitude hum trace should be displayed.

To avoid the instrument responding to hum signals from the mains circuit and heater line, it is advisable to use screened Y-input cable for this test. As much Y gain as possible should be used.

Under very sensitive conditions, slight ripple will be seen when the Y input is connected to most valve cathodes, but a faulty valve will produce a display of far greater amplitude. On a valve that has a large electrolytic capacitor decoupling the cathode circuit much of the hum signal will be bypassed, but there is generally sufficient remaining to show when a valve is faulty. Of course, for the heater-cathode leakage to inject hum signal into the associated circuits there must be a certain amplitude of ripple voltage across the cathode resistor.

**Capacitor check**

The scope can be used to check the efficiency of bypass and decoupling components. For instance the electrolytic across, say, the cathode resistor of the sound output valve should offer a low impedance to all but the very lowest audio signals. If the capacitor is in good order no signal voltage should exist across the cathode resistor. If a scope test on the cathode of the valve reveals the presence of an audio signal (Fig. 6.1) when the sound channel is working, the electrolytic capacitor should be changed.

The same reasoning applies to all bypass capacitors in the r.f., i.f. and a.f. stages. At the higher frequencies this test is not possible unless the scope's Y amplifier responds to such signals.

This sort of testing can be extended into the timebase circuits to some degree but it must be borne in mind that while some capacitors may appear at first sight to be concerned solely with bypassing, their purpose may be to provide feedback for scan linearization and so forth.

In the video amplifier stage, the cathode bypass capacitor is rarely intended to provide a low impedance to the very lowest of video frequencies (this is

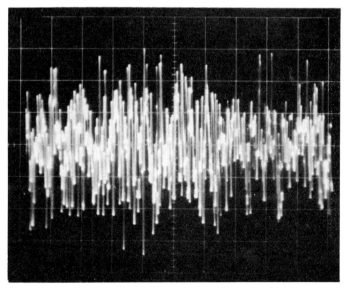

*Fig. 6.1.* This is how audio signal appears at the cathode of the sound amplifier when the bypass capacitor is open-circuit.

impossible, anyway, since these theoretically go down to d.c.). While the higher frequency signals are bypassed, the lower ones appear across the cathode resistor, and these are often purposely arranged (as negative feedback) to control the overall response of the stage – in conjunction with other compensating networks in the anode circuit and in the feed to the picture tube.

#### Use of signal generator

While off-the-air sound and vision signals, and signals generated in the set itself, provide the majority of waveforms for scope testing, there are times when an externally generated sine or square-wave signal is desirable. This signal can be applied to the stage or network under test, and the output compared on the screen of the scope with the known 'shape' of the input signal.

For this purpose a sine and square-wave signal generator covering about 20 Hz to 1 MHz is a very useful instrument to partner the oscilloscope. One of the best-known methods of assessing the performance of an audio or video amplifier is to feed into it a good square-wave signal and check the characteristics of the waveform seen on a scope connected across the output load. Any deviation from the true square-wave form can be used to judge the design, and fault condition (if any) of the amplifier.

The signal should be applied to the control grid of the video amplifier via a 0·25 µF capacitor with the Y input of the scope connected via a low-loss high-impedance coupling to the anode or output circuit of the video amplifier (i.e. the tube cathode feed). In practice it is best to remove the socket from the tube base (restoring heater chain continuity with a series resistor across the heater tags) in case there is any slight leakage from the tube cathode. This would impair the output waveform display.

This effect can, in fact, be used to assess whether the tube is responsible for any shortcoming in video response. The idea is to compare the square-wave output waveform as obtained with the tube base removed and with it connected. If the tube is in good condition, there should be only the very smallest trace of corner rounding of the square wave with the tube connected. When the tube is actually working the leakage may alter.

#### Square-wave testing in video amplifier

A square wave is a good testing signal because it is composed of a large number of harmonically and phase related sine-wave signals (see Chapter 1). Thus if the amplifier and circuits are able to pass the square wave without distortion, one can be sure that component signals are not being lost or changed in phase in the amplifier.

The fundamental frequency and a series of odd-numbered harmonics produce the square wave. The waveform becomes more rectangular as more

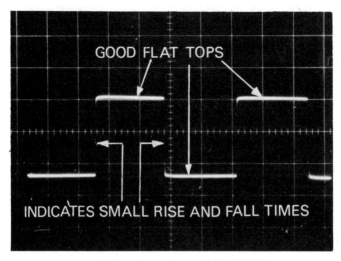

*Fig. 6.2.* Example of good 10 kHz display.

odd-numbered harmonics are added to the fundamental frequency. The accuracy of the square wave, however, is also associated with the relative amplitudes and phase relationships of the component frequencies, so it can be appreciated why a square wave is a critical test signal. Any change in phase or amplitude of component frequencies will alter the shape of the waveform at the output. Moreover, the fast rise and fall produce ringing effects in any amplifier prone to ringing and overshoot.

To preserve the square nature of the waveform, an amplifier must have a bandwidth up to at least ten times the square-wave repetition frequency. This means that the video amplifier should be capable of passing a square wave of at least 500 kHz, but the square-wave nature of the output waveform may be changed by the frequency compensation applied to some video amplifiers, including the primary and colour-difference amplifiers of colour receivers. A well designed video channel may pass square waves in excess of this frequency, for the rate of roll-off outside the band may not be all that rapid except when a filter lies in the circuit to minimise sound/vision carrier beats (at 3·5 MHz) or intercarrier beats (6 MHz).

A good square wave from the output of a video amplifier at 10 kHz is shown in Fig. 6.2. Increased rounding of the corners was evidenced at higher frequencies, as shown in Fig. 6.3. Fig. 6.4 reveals even more rounding as the generator frequency is further increased. To avoid false results due to excessive shunt capacitance from the Y input connecting circuit it is important to employ a correctly compensated 10:1 low-capacitance probe for h.f. square wave displays.

SERVICING WITH THE OSCILLOSCOPE

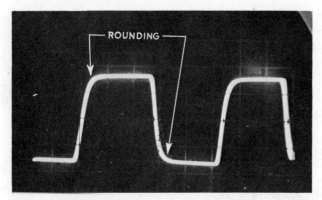

*Fig. 6.3.* Reduction in h.f. response is indicated by corner rounding, as shown in this oscillogram.

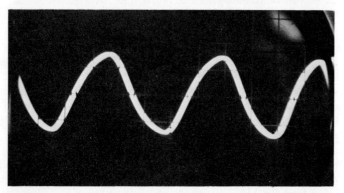

*Fig. 6.4.* When the bandwidth is limited to the repetition frequency of the square wave, the square wave nature is destroyed and the display appears like a distorted sine wave.

This degree of rounding may not particularly signify a video amplifier fault. The aim should be for the amplifier to pass a 100 kHz square signal with little more rounding and reduction in rise time than shown in waveform Fig. 6.5. Indeed, this would be quite a good characteristic.

However, if the rounding were anything like that of Fig. 6.4 at 100 kHz there would be room for improvement. Causes of trouble could be increase in value of the anode load resistor in the video amplifier, excessive capacitance at the coupling from the video amplifier anode circuit to the picture tube and excessive reactive loading in the picture-tube gun assembly.

# TELEVISION TESTS FOR HUM, DISTORTION AND RESPONSE

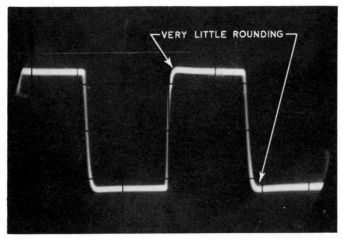

*Fig. 6.5.* Video amplifier square wave response at 100 kHz.

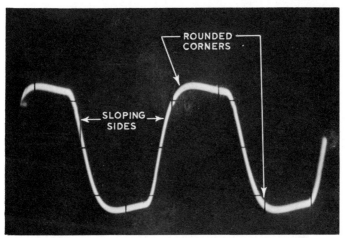

*Fig. 6.6.* Square-wave input approaching sine wave characteristics at the output. This was obtained at the cathode of the picture tube with parallel capacitance added.

The latter can be tested by running with the cathode feed to the picture tube disconnected at the video amplifier valve anode. (Make sure the tube is not severely under-biased by this action. It is best to remove the grid connection and the e.h.t. to the final anode when making tests of this kind.)

Fig. 6.6 shows the display obtained with a little capacitance at the cathode

SERVICING WITH THE OSCILLOSCOPE

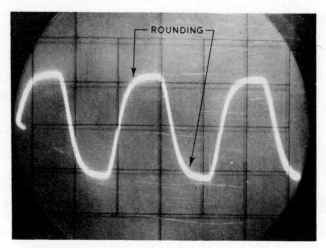

*Fig. 6.7.* Waveform obtained at video amplifier anode when the load was increased to 100 k (from about 2·7 k). Frequency 100 kHz.

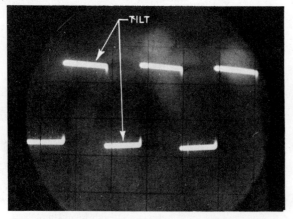

*Fig. 6.8.* The tilt here is caused by very slightly falling l.f. response. Frequency 50 Hz.

feed to the picture tube and Fig. 6.7 shows what happens when the anode load of the video amplifier valve goes up to 100 k or thereabouts. The square wave in Fig. 6.6 has virtually changed into a sine wave, though rather a distorted one. It implies that the amplifier is capable of passing only the fundamental frequency of the signal.

Revealing tests can also be made at lower square-wave repetition frequencies. For instance, a test at round 50 Hz will show the low-frequency performance of the amplifier, and as the response in many sets is well down at that frequency a substantial distortion of the waveform is to be expected.

Fig. 6.8 shows a typical display at low-frequency, the tilting being caused by a relatively poor l.f. response, though not to an abnormal degree. The tilt indicates that the amplifier has a leading low-frequency phase shift. Another waveform indicating l.f. attenuation is given in Fig. 6.9.

The waveforms shown in Figs. 6.10 and 6.11 indicate a fairly good l.f. response with substantial overshoot characteristic. This could be responsible for black smudges following white picture content. A waveform something like that given in Fig. 6.12 may be obtained at an intermediate frequency. This implies relative h.f. attenuation with severe l.f. overshoot (ringing).

The extended h.f. bandwidth of a video amplifier is achieved by the use of a relatively low-value output load resistor and a high-slope valve or high-gain and high-$f_T$ (gain-bandwidth product in common-emitter mode) transistor. The idea is to ensure that the shunting effect on the h.f. signal components due to stray and circuit capacitances is small in terms of the shunt-impedance to load-impedance ratio. So-called peaking coils are often used to maintain the response to take over at the frequency where the gain of the stage tends to fall off.

Square-wave testing in the video amplifier may thus show ringing due to the inductive frequency-compensating components. A typical ring (damped

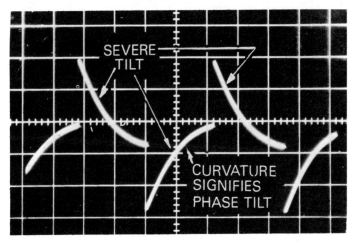

*Fig. 6.9.* Another display of l.f. attenuation and mild phase shift.

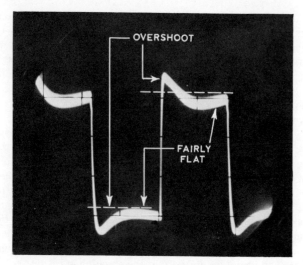

*Fig. 6.10.* Overshoot on waveform.

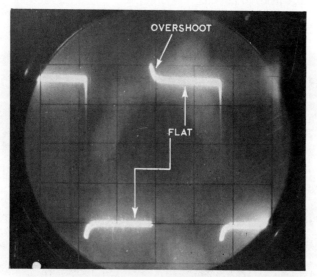

*Fig. 6.11.* Another display of overshoot, but this time with good l.f. characteristics. Overshoot in video amplifier stages can cause black smudges from white picture content.

# TELEVISION TESTS FOR HUM, DISTORTION AND RESPONSE

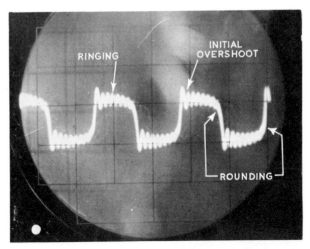

*Fig. 6.12.* L.f. overshoot and h.f. ringing.

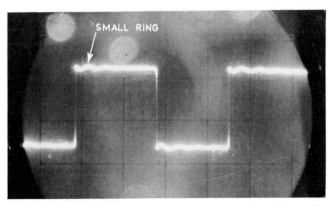

*Fig. 6.13.* Another ringing waveform. Rings like this (and in Fig. 6.12) cause vertical bars on picture.

oscillation) evoked by a sudden change in signal amplitude is shown in Fig. 6.13. This magnitude of ringing could not generally be tolerated as it would cause alternate dark and light vertical lines across the picture. The ringing is damped by resistors connected across the inductive elements, thereby lowering their $Q$.

It is possible to appraise the overall square-wave performance of the vision detector and video amplifier, but to do this correctly the square wave should be modulated on to a carrier-wave at the vision i.f. This modulated signal may then be applied to the input of the vision detector (or further along the i.f. channel towards the tuner) and the demodulated square-wave signal analysed along the line already described. The signal generator, of course, must have a means of accepting external modulation over a wide modulation bandwidth and there are few servicing-type instruments that are capable of this facility.

Square-wave signals can be used to determine the operation of the sync separator circuits and differentiator and integrator feeds to the line and field generators, but in the main it is best to employ the actual signal sync pulses for this purpose if at all possible.

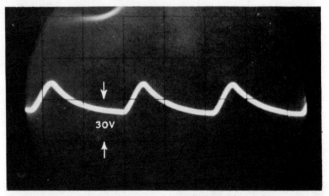

*Fig. 6.14.* Ripple voltage across the reservoir capacitor.

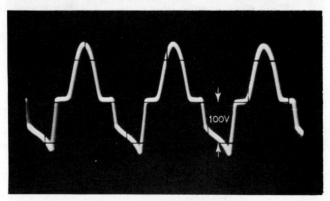

*Fig. 6.15.* Waveform at centre of a pair of series-connected h.t. rectifiers.

# TELEVISION TESTS FOR HUM, DISTORTION AND RESPONSE

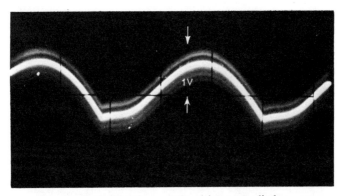

*Fig. 6.16.* Waveform on the vision a.g.c. diode.

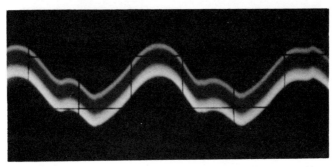

*Fig. 6.17.* Signal on vision a.g.c. line. The amplitude of this signal varies with changes in picture and with setting of contrast control.

**Miscellaneous tests**

To conclude this chapter we will look at one or two other oscilloscope tests that can help with fault diagnosis.

Fig. 6.14 shows the ripple voltage across the reservoir capacitor at the positive side of the metal h.t. rectifier. The sweep was adjusted to give about two cycles of ripple waveform and the Y input to 30 V/div. The ripple amplitude at this point is just about correct. It should be considerably less (barely detectable) on the h.t. line proper, that is after the first smoothing stage, across the main smoothing capacitor.

Fig. 6.15 is an interesting oscillogram of the waveform at the centre of a pair of series-connected h.t. rectifiers. This shows the symmetrical operation of the rectifier on each half cycle. The sweep is again adjusted to display the ripple waveform, while the Y input is adjusted to 100 V/div.

SERVICING WITH THE OSCILLOSCOPE

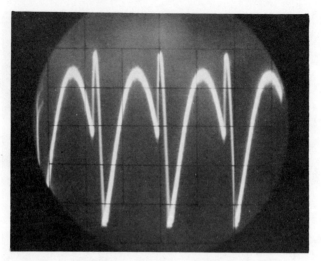

*Fig. 6.18.* Signal on focus electrode of picture tube.

Fig. 6.16 shows the waveform present on a vision a.g.c. diode. The sweep is set at 6 ms/div and Y input to 1 V/cm. The waveform shows some integrated field signal coupled with some mains hum, along with some residual line signal. This latter signal is shown by the thick white line that trails along the waveform. A similar waveform is shown in Fig. 6.17 but this time it is from the actual vision a.g.c. line feeding the controlled valves. The amplitude of this signal varies with changes in picture content and with the setting of the contrast control.

The signal on the focus electrode of the picture tube is seen in Fig. 6.18. Sweep is set towards 30 µs/cm and Y input to 3 V/cm and the waveform results from the fact that the focus electrode derives its potential partly from the boosted h.t. supply on which line signal is often superimposed, depending on the nature of the circuit.

# 7: VISUAL CIRCUIT ALIGNMENT

THE oscilloscope can also be used to display the response characteristics of tuned amplifiers, which is an extremely useful service because it enables the engineer to see exactly the effects that the tuning cores have on the response curve as they are adjusted.

For this display the alignment input signal to the receiver's circuit must be swept over the whole frequency range of the response. The 'sweep' must be in synchronism with the X sweep of the spot on the screen of the scope. For television, the Y input of the scope is best connected to the sound or vision detector output, so that vertical (Y deflection) of the spot corresponds to the level of the signal at the output of the tuned channel. Since the level of signal changes over the pass band, the vertical deflection changes accordingly and, since the X sweep is 'locked' to the frequency sweep, it follows that the response characteristic is traced on the screen.

Set-up of instruments and their connections to a television receiver are shown in Fig. 7.1. Here the wobbulator supplies the input alignment signal to the receiver and the response at the detector's output gives Y deflection

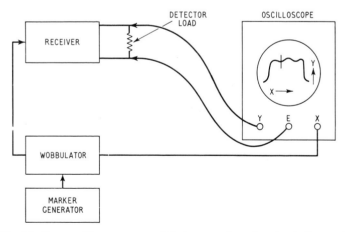

*Fig. 7.1.* Set-up of instruments and their connections for visual alignment.

115

# SERVICING WITH THE OSCILLOSCOPE

on the oscilloscope. It will be seen that d.c. coupling is used from the detector load to the Y input so that the vertical deflection corresponds to the d.c. voltage across the load.

### How the response curve is formed

Let us consider rather more closely how the system operates. In the first place we know that the rectified voltage (d.c.) at a detector rises from zero with no signal input to a peak reached when the input signal is at the nominal or centre tuned frequency of the amplifier, and then falls again as the input signal passes to and beyond the upper limit of the response band. Also, it follows that if the detector output is connected to the Y plates of a scope, with the timebase switched off, the spot will rise and fall vertically in the middle of the screen as the input signal is applied to the tuned amplifier and swept through the pass band.

This up and down movement would tell the observer very little about the 'shape' of the response – whether it was flattish or peaky or double-humped

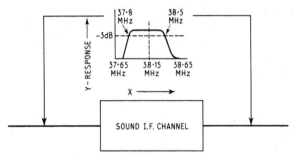

*Fig. 7.2.* The relative response of a sound channel as may be traced out with a sweep of about ±500 kHz.

and so forth – unless he laboriously injected a series of spot frequencies and measured the corresponding heights of the spot. The value of the wobbulator system is that the sweep of the signal through the frequency band is synchronised with the sweep of the spot and repeated, so that the variations in response are seen as a curve.

The action can be obtained by using the same control voltage to vary the wobbulator's frequency and to deflect the scope's spot. Generally, the X voltage of the oscilloscope's timebase is used for the purpose, this simply requiring a connection from the scope to the wobbulator.

The wobbulator embodies an electronic reactance or a ferrite modulator across the tuned circuit of its oscillator. The value of this is made to change linearly with rise in X voltage from the scope's timebase. The frequency of

## VISUAL CIRCUIT ALIGNMENT

the wobbulator is swung from a negative to a positive value relative to the nominal, tuned frequency. So the position of the spot on the oscilloscope screen corresponds horizontally to frequency and vertically to detector output.

Fig. 7.2 shows the relative response of a sound channel as may be traced out with a sweep of about $\pm 500$ kHz.

When the wobbulator itself supplies the scope sweep voltage, the 50 Hz mains frequency is often used to sweep the frequency and made available,

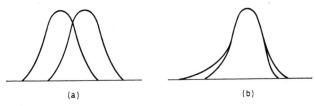

(a)                    (b)

*Fig. 7.3.* When 50 Hz sine wave sweep is used, the two half-cycles of sweep signal will produce two traces side by side as at (*a*). These are made coincident as at (*b*) by means of a phasing circuit.

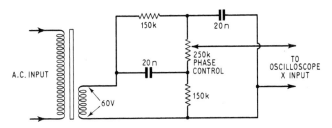

*Fig. 7.4.* A phasing circuit to obtain the condition illustrated in Fig. 7.3(*b*).

via a phasing network and control, at a terminal for connecting to the scope's X input for horizontal deflection – the scope's internal timebase then being switched off, of course.

The phasing control is required because the sweep is derived from a near sine wave (the mains waveform) which results in display on both the forward and return traces. The phasing control ensures that the two traces coincide on the screen (see Fig. 7.3). The need for this is avoided in some instruments by automatic blanking of the return trace. A typical phasing control network is shown in Fig. 7.4.

## Marker pip

Fig. 7.1 shows an instrument called a marker generator, which has not yet been mentioned. This plays no part in the actual creation of the response display and is employed so that points on the display may be accurately 'placed' in frequency. For instance, in Fig. 7.2 the exact frequencies at which the response is 3 dB down (i.e. the half-power points) need to be known. The wobbulator itself cannot determine these and so a marker generator is used. This can be any ordinary signal generator of reasonable accuracy, used with the modulation switched off.

The marker signal can be injected into some wobbulators at a signal generator terminal. Even if no such input connection is provided it is not difficult to inject the signal. One way is to wind a couple of turns of wire from the marker generator round the signal lead from the wobbulator. The set then receives the wobbulated (or swept) signal and a little of the marker signal, and the beat produced appears as a 'pip' on the trace as shown in Fig. 7.5. Too much marker signal can distort the trace pretty badly and make correct alignment almost impossible.

## Sweep techniques

Maximum sweep is determined by the nature of the particular instrument. In most instruments, sweep up to the maximum from zero is obtainable by a control. The sweep control usually regulates the level of signal applied to the frequency modulator of the wobbulator.

It has already been mentioned that an electronic device is used to swing the oscillator tuned circuit. Another way employs a capacitance diode (varicap) across the oscillator circuit. The capacitance of this semiconductor device changes as the reverse bias applied to it is altered. The reverse bias corresponds to the sweep control voltage picked up either from the scope's timebase or from the mains supply.

Television receiver alignment (of the vision channel) requires a wide-range linear sweep. This is sometimes obtained from a ferrite modulator, which consists of a ferrite core carrying an r.f. winding forming a part of the oscillator tuned circuit. The permeability of the core is set to a nominal value by a permanent magnet. The core is mounted in the gap of a different type of core, upon which is wound a coil carrying the sweep current. The sweep current causes a change in the permeability of the ferrite core, which is reflected as a change in the inductance of the r.f. winding, and, consequently, a change in the tuned frequency.

## Alignment techniques

Wobbulators for radio and television sound-channel alignment, including f.m. channels, need to sweep over about $\pm 400$ kHz, but a sweep up to 6

## VISUAL CIRCUIT ALIGNMENT

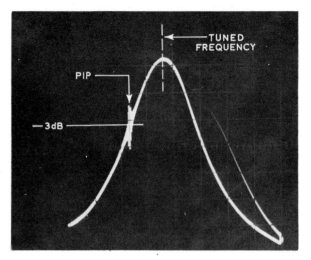

*Fig. 7.5.* The formation of a 'pip' on a response trace.

MHz or so is required for vision channels.

The frequency of sweep is, of course, fixed at 50 Hz when wobbulator control is by the mains frequency; but when the scope's timebase is used it is possible to sweep at any frequency provided by the scope. This is where real trouble can start, for if too high a sweep frequency is adopted a distorted response curve can be made to look quite linear and smooth. On the other hand, if too low a frequency is adopted, not only will flicker make it difficult to study the curve display but distortion will be aggravated on the response.

The best plan is to set the timebase sweep to about 30 Hz, just a little above bad flicker, and make sure that this frequency is retained throughout the whole of the alignment exercise.

When the Y signal is obtained from the detector load it may be necessary to provide a little Y gain. To avoid trace distortion the Y amplifier should respond well at d.c.

Vertical deflection can be increased both by increasing the alignment input signal to the set (from the wobbulator) and the Y gain on the scope. Again, this is where trouble can be introduced. A large alignment signal and small Y gain can cause overloading in the receiver circuits and a consequent artificial flattening of the top of the response curve. Conversely, a small alignment signal and large Y gain can result in hum problems and grass (noise) on the top of the response display.

The author favours adjusting the alignment-signal level and Y gain first to give a display with a little grass on top, and then increasing the input signal and decreasing the Y gain until the grass just disappears, aiming for about two-thirds screen amplitude in the Y direction.

If hum causes pulsating of the trace on the screen one can reduce the Y gain and increase the alignment signal, or try to lock the hum by critically adjusting the scope sweep frequency, or to reduce the hum pick-up by using a screened Y lead between scope and receiver.

### Marker distortion

Since there is a possibility of the marker signal distorting the response display it is a good idea to switch it off from time to time during the alignment process or, better still, to switch the marker on only when it is required to identify frequency on the curve. The marker pip can sometimes be better defined on the trace by connecting a 0·01 µF capacitor across the scope Y and earth terminals.

### Input to receiver

The wobbulator signal should be applied to the required input point of the receiver through the normal resistive padding system, as dictated by the circuit. If the response right from the tuner is to be displayed, the signal should be applied, of course, to the aerial socket across the correct impedance. A low-impedance connection to the input of the i.f. amplifier is also often permissible. Where the set is connected direct to the mains supply (a.c./d.c. type), as is nearly always the case, capacitor isolation must be adopted or, to avoid the hum problems that this kind of isolation can incite, the set energised from a 1:1 ratio mains isolating transformer. The instruments can then be coupled in direct without danger of a destroying mains loop.

### Receiver output

The signal for the scope curve can be obtained either before or after the detector. To obtain a pre-detector display, an external detector network must be used, such as that shown in Fig. 7.6. The signal a little way down the amplifier chain will be at lower level than at the detector and an external detector output has to be coupled to a relatively high Y gain input on the scope to obtain sufficient vertical deflection. Screening then becomes particularly important to prevent hum on the display.

The Y signal can also be obtained from the output of the first a.f. amplifier or from the output of the video amplifier in the vision channel (i.e. from the cathode of the picture tube). This is possible because the Y signal is effectively modulated at the sweep frequency, and both audio and video amplifiers should pass sufficient signal at that frequency.

# VISUAL CIRCUIT ALIGNMENT

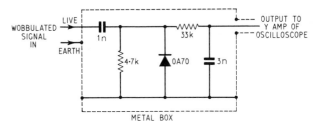

*Fig. 7.6.* An external detector circuit, as shown here, is used to obtain a pre-detector response trace.

## General hints

There is no point in elaborately detailing the alignment procedures for radio and television sets for these are clearly given in service manuals. One or two visual alignment hints will not be amiss, however.

It is best to apply the input signal to the first i.f. stage or to the tuner i.f. test point and to take the output from the detector load, as in Fig. 7.1. When the vision response trace is obtained, the marker generator can be set to the carrier frequency and the cores of the transformers adjusted carefully until the response indicated in the manual is obtained, making sure that the carrier remains at the correct point down the response for single-sideband working. The response on both 405-lines and 625-lines should be checked and, if necessary, corrected on dual-standard sets.

## Rejector adjustments

The marker generator should then be accurately tuned to the sound i.f. and the sound rejectors adjusted to give the deepest trough in the response at that frequency. Adjacent-channel and other response-tailoring rejectors should be adjusted in similar manner, the frequency of each one being identified on the response display by means of the marker generator.

This method of alignment often avoids the need for damping components across one side of a transformer while the other side is being adjusted, as is necessary with point-frequency alignment.

It is possible to stretch out any part of the response display either by reducing the frequency range of sweep or by expanding the scope X sweep by the X-gain control. This technique is often desirable when a rejector is being adjusted to ensure that the maximum possible attenuation at its tuned frequency is obtained. A greater alignment signal input can be applied to lift the response so that its greatly attenuated parts are increased in amplitude on the screen.

It is important to ensure that the sound rejector alignment frequency in the

# SERVICING WITH THE OSCILLOSCOPE

vision channel is exactly the same as that used for aligning the sound channel. A dual-trace scope can assist in this connection by displaying the vision response on one trace and the sound response on the other. This simply entails connecting one Y input across the vision detector output and the other Y input across the sound detector output.

### Sound channel alignment

Fig. 7.7 shows the response of a sound channel with the marker pip identifying exactly the nominal sound i.f. In Fig. 7.5 the marker pip is shifted down the leading side of the response (about 3 dB down) and in Fig. 7.8 the marker pip is about 3 dB down to the trailing side. The bandwidth of the sound channel is equal to the marker frequency in Fig. 7.8, minus the marker frequency in Fig. 7.5. Adjustments to the response would be made (in the 405-line sound channel) to give the conditions shown in Fig. 7.2.

On 625 lines it is very important to ensure that the intercarrier channel is aligned exactly to 6 MHz. This is best achieved by applying a 6 MHz nominal wobbulator signal to the input of the sound i.f. strip and observing the overall response at the anode of the final vision i.f. transformer with an external detector (Fig. 7.6). The marker generator should then be very carefully adjusted to deliver a 6 MHz signal. The cores of the 6 MHz tuned circuits can next be adjusted to give the display shown in Fig. 7.7, the aim being for the marker pip to fall exactly at the top of the response and for the response to fall to the 3 dB points at about $\pm 40$ kHz. Too narrow a response here can exaggerate sound channel drift and response distortion in the f.m. detector while too wide a bandwidth could lead to interference troubles (intercarrier buzz, etc.).

On the 405-line sound channel too wide a response will give bad vision-on-sound and too narrow a response may lead to critical adjustment of the

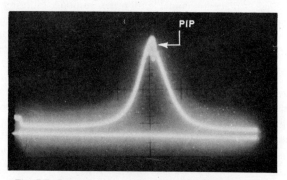

*Fig. 7.7.* Response trace of sound channel with marker pip showing exact tuned frequency.

## VISUAL CIRCUIT ALIGNMENT

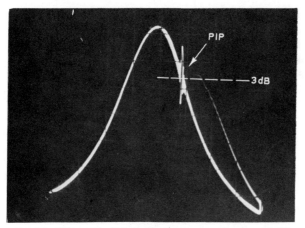

*Fig. 7.8.* Where Fig. 7.5 shows the marker pip about 3 dB down the leading side of response, this oscillogram shows the pip about 3 dB down the trailing side of the curve.

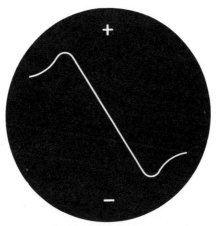

*Fig. 7.9.* Symmetrical f.m. detector response characteristic.

fine tuning control and sound i.f. instability, coupled with poor interference limiting.

When the intercarrier tuning has been finalized, the external detector should be disconnected and the output of the f.m. detector applied to the Y input of the scope. The ratio detector transformer should then be adjusted to give a symmetrical response as in Fig. 7.9. This shows a really good response for this type of circuit.

123

A more practical response here is seen in Fig. 7.10. The aim has been to obtain a linear diagonal part to the characteristic, from the rising positive peak on the left to the falling negative peak on the right. If a marker pip is now applied, this should appear approximately in the centre of the diagonal part of the trace, at the intersection of the X and Y lines of the graticule in Fig. 7.10.

After f.m. alignment the scope's Y input can be transferred to the audio amplifier output of the sound channel and the alignment signal frequency modulated with a tone. Some wobbulators incorporate such a facility, the modulation being provided by an inbuilt 1,000 Hz sine-wave generator.

Fig. 7.11 shows the audio waveform when the f.m. signal is detuned a little from the intercarrier frequency or, conversely, when the intercarrier channel is not correctly aligned to 6 MHz. Correct alignment is signified by a reasonably pure sine-wave display (Fig. 7.12). This waveform however is composed of about 5 per cent second-harmonic distortion (signified by the rounding of the negative half-cycles relative to the positive half-cycles). This level of distortion is permissible in domestic television sets running towards full audio output.

**Distortion**

Fig. 7.13 shows the effect of clipping in the first audio amplifier of the sound channel (after the detector) due either to overloading (too much input

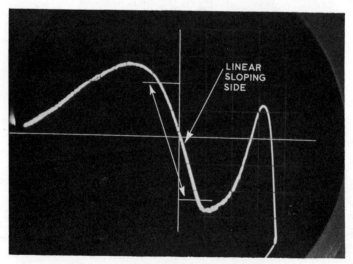

*Fig. 7.10.* A more practical f.m. detector response curve.

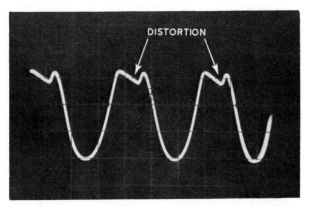

*Fig. 7.11.* Audio signal with mistuned f.m. detector. Sine wave input signal.

signal) or poor bias adjustment. A low-emission valve could give a similar effect.

Figs. 7.14 and 7.15 give some idea of how the response trace may appear during the alignment exercise, with the second trace of a dual-trace scope acting as a base line.

Fig. 7.16 shows how the vision channel response may appear with the tuned circuits badly out of alignment. The two peaks are caused by two tuned circuits in the channel responding individually. As they are brought into correct alignment the two peaks move towards each other and eventually combine to give a wideband, composite vision response curve.

The nature of the vision response curve when the circuits are in correct alignment is usually revealed on the circuit or in the alignment instructions. The rejector and carrier frequencies are also shown, and it is very important to continue with the adjustments until the scope display coincides as closely as possible with that shown in the maker's manual.

**Importance in colour sets**

One gradually becomes more proficient with this method of receiver alignment. Although the first three or four attempts will probably take longer than the spot-frequency method, time will be cut dramatically with experience. The net result is a far better alignment.

The importance of very accurate alignment is proved with dual-standard monochrome sets, but with colour sets absolute alignment, aided by the visual procedure, is essential, even though the remainder of servicing may resolve to the changing of modules! Poor alignment in colour sets will attenuate the subcarrier and the colour modulation components – the same as a television aerial with insufficient bandwidth.

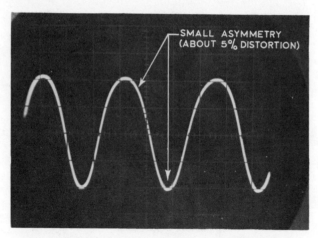

*Fig. 7.12.* When the f.m. detector is correctly tuned this waveform results (from sine wave input). Here there is about 5 per cent harmonic distortion.

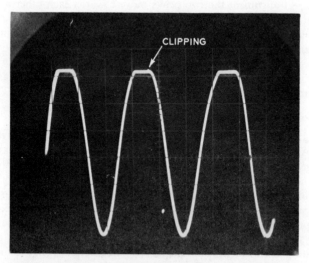

*Fig. 7.13.* Asymmetrical clipping in first audio amplifier.

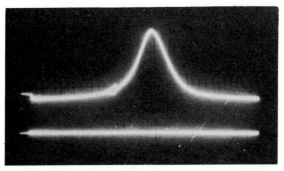

*Fig. 7.14.* Showing how the response trace may change during alignment.

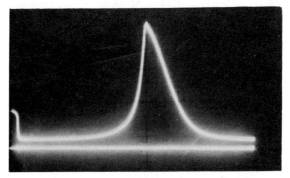

*Fig. 7.15.* Another appearance of the trace during alignment.

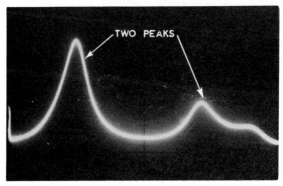

*Fig. 7.16.* There may be a tendency for two peaks to occur when the alignment is very bad.

## Application to other tuned circuits

Visual alignment techniques can be applied to any tuned amplifier or tuned channel, but amplifiers and equipment without a detector must have the 'swept' signal demodulated before application to the Y input of the scope. Nothing elaborate is required here, and a simple diode detector circuit with the usual load and reservoir capacitor is adequate. The signal from across the detector load is applied to the Y input of the scope. If the demodulated signal is fairly weak, a substantial Y gain may be needed to secure sufficient vertical deflection of the trace, but the Y bandwidth is not too important in this application since the signal is swept at low frequency and the amplifier is only handling signal changing at that rate. The l.f. response is more important, really, than the h.f. response of the Y amplifier, for there must be reasonable Y amplification at the relatively low sweep frequency. This does not present radio application problems, for not a great deal of Y gain is needed to get the required vertical deflection of the trace. However, when an external Y amplifier has to be interposed to lift a very low-level detector signal to a suitable Y input value, due attention must be given to the amplifier's l.f. gain.

Any ordinary a.m. and f.m. radio set can be aligned visually by adopting the techniques already expounded, and by using marker pips the overall bandwidth required for any application can be guaranteed. This is particularly important for f.m. tuners and radios incorporating a stereo decoder, as explained in Chapter 9. A dual-sweep function, which also determines f.m. intermodulation distortion, is provided by Sound Technology Type 1000A stereo f.m. generator.

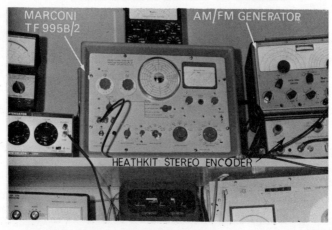

*Fig. 7.17.* Part of the author's laboratory, showing the Marconi TF995B/2 v.h.f. f.m./a.m. signal generator and various other instruments.

## VISUAL CIRCUIT ALIGNMENT

**F. M. tuner alignment**

Much of what has already been said is applicable also to hi-fi f.m. tuners. In general, however, equipment of a magnitude or so better than that required for normal servicing applications is essential for detailed evaluation of the hi-fi type of f.m. tuner.

One item of equipment used by the author for visual response measurements is the Marconi TF995B/2 a.m./f.m. generator. This has external modulation facilities, thereby making it possible to sweep from the X (timebase) output of an oscilloscope. This instrument, with some others in the author's laboratory, is shown in Fig. 7.17.

The i.f. channel bandwidth of a correctly limited f.m. tuner increases as the input signal level increases which, owing to the amplitude limiting effect of the tuner, is tantamount to measuring the bandwidth at various levels down the response characteristic. The series of sweep oscillograms in Fig. 7.18 illustrates this effect. At (a) is shown a sweep response with a signal input insufficient to take the tuner into limiting (hence the noise along the top of the characteristic). At (b) the signal level is increased which brings the tuner into limiting, thereby deleting the noise along the top. At (c) an even greater level of input signal is applied, which pushes the limiting down the response further.

Thus, if the horizontal divisions are calibrated in frequency per division, the bandwidth at any level down the response characteristic can be determined. All the oscillograms in Fig. 7.18 are calibrated so that each division horizontally corresponds to 80 kHz, which puts the bandwidth at (a) about 200 kHz, at (b) about 240 kHz and at (c) about 480 kHz. From such a series of responses, therefore, it is possible to calculate the rate of slope of the side skirts of the response characteristic, as shown in Fig. 7.19.

For good quality stereo reproduction from f.m. radio the bandwidth at −6 dB should be around 240 kHz – not much less – and the skirts of the response should be as steep as possible (provided in the latest f.m. tuners by ceramic filters) thereafter.

The frequency per horizontal division is best calibrated by means of a digital frequency counter. The counter is used to establish the mean frequency of the input signal by lining up the display with the start of the first division (say, on the left-hand side of the graticule) by shifting the frequency of the generator and hence the display, then measuring this frequency with the counter. The display is then shifted so that the same point of the display lines up with the finish of the final division (on the left-hand side of the graticule), and this frequency is then measured (note that when the swept-generator frequency is changed the display moves across the screen, in a direction depending on whether the change is up or down). The difference between the two measured frequencies gives the total bandwidth,

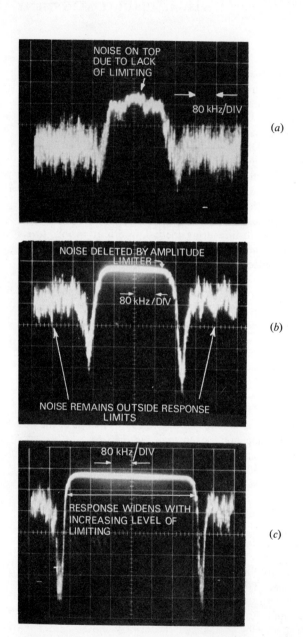

*Fig. 7.18.* F.M. tuner response oscillograms at various signal levels: (*a*) just prior to limiting; (*b*) at about 20 dB into limiting; (*c*) at about 50 dB into limiting.

## VISUAL CIRCUIT ALIGNMENT

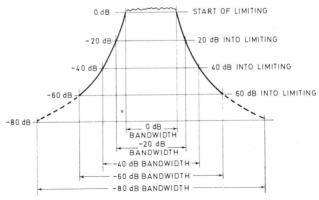

*Fig. 7.19.* Showing how the rate of slope of the response side skirts can be calculated or appraised by measuring the bandwidth at various levels into limiting (see text). This technique also reveals any asymmetry of response characteristic.

and if this embraces, say, ten horizontal divisions the frequency per graticule division can be calculated. For example, if the total shift over ten divisions works out to 500 kHz, then the frequency per division is 500/10, or 50 kHz, assuming a linear (ramp) sweep.

An accurate, though not particularly costly, frequency counter is the Heathkit IB–1101 shown in Fig. 7.20. This reads v.h.f. signals up to 100 MHz and low-frequency signals down to 1 Hz. A very useful acquisition.

*Fig. 7.20.* Model IB–1101 v.h.f. frequency counter by Heathkit.

# 8: COLOUR TELEVISION WAVEFORMS

MANY of the waveforms illustrated in the earlier chapters for monochrome television are present also in colour receivers. However, colour sets embody circuits specifically designed to process the colour signals and to control the colour picture tube, and these produce their own waveforms.

The scene at the studio or other programme source is analysed into its three additive primary colours, red ($R$), green ($G$) and blue ($B$), and the source delivers signals corresponding to these three colours in the proportions $R$30 per cent, $G$59 per cent and $B$11 per cent relative to 'white' input. These proportions correspond to the three primary components of the picture in monochrome. Thus, their addition corresponds to the monochrome signal, as would be delivered by a *monochrome* camera scanning the same scene. The monochrome signal in colour parlance is called the *luminance* or $Y$ signal.

Colour television systems are compatible, meaning that colour-encoded signals will work ordinary monochrome sets (in black-and-white), while colour sets will also respond in black-and-white to ordinary monochrome transmissions. The $Y$ signal gives the compatibility, for this is modulated upon the carrier in the ordinary way, while the $R$, $G$ and $B$ signals are integrated upon the same modulated carrier in such a manner that their effect is negligible on monochrome sets, while contributing the actual colour on colour sets.

### Colour difference signals

Colour-difference signals are created at the transmitter by a matrixing network which subtracts the Y signal separately from the $R$, $G$ and $B$ signals. The signals remaining are then $(R-Y), (G-Y)$ and $(B-Y)$. Since the $Y$ signal must be applied to the transmitter in the ordinary way (for monochrome sets receiving the transmission), only two colour-difference signals need to be encoded, and those chosen are the $(R-Y)$ and $(B-Y)$ ones. The $(G-Y)$ signal is retrieved at the receiver by matrixing the other two.

The red and blue colour-difference signals are amplitude-modulated upon a 4·43361875 MHz subcarrier at the transmitter in quadrature. This

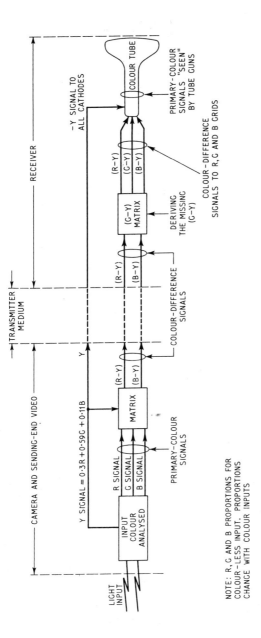

*Fig. 8.1.* Basic colour television system where colour-difference drive is used at the display tube in the receiver. The alternative, now commonly used, is primary colour or *RGB* drive to the tube guns. With this arrangement the primary colour signals are obtained prior to the tube by a process of matrixing whereby *Y* signal is added to each of the colour-difference signals. Thus $R = (R-Y)+Y$, $G = (G-Y)+Y$ and $B = (B-Y)+Y$.

frequency is accurately chosen to avoid interference. Prior to transmission the subcarrier is suppressed, leaving only the upper and lower sidebands which are added to the $Y$ modulation at the main carrier.

**Basic principles**

The picture tube of a colour set responds separately to the colour signals and to the $Y$ signal. Thus, when such a tube is fed only with $Y$ signal (grid voltages set for white light) it gives an output in black-and-white, while when fed simultaneously with $Y$ plus $R$, $G$ and $B$ colour-difference signals it gives an output in colour.

The colour tube has three separate electron guns, and hence three electron beams. Each gun, one for each primary colour, is very similar to the single gun in a monochrome picture tube. The cathodes of the three guns are generally fed together with $Y$ signal, giving the luminance or monochrome presentation, while the grids are separately fed with $(R-Y)$, $(G-Y)$ and $(B-Y)$ signals. The very basic arrangement is shown in Fig. 8.1.

The three-gun tube has a special kind of screen composed of a multiplicity of phosphor dots, instead of the even phosphor distribution of monochrome tubes, and the dots are arranged in groups of three, with each one of a group having the property of glowing red, green and blue individually. The three electron guns are so placed in the tube neck that after being accurately adjusted by fixed and dynamic magnetic convergence fields their beams impinge only upon the corresponding colour phosphors, and this coincidence is retained even when the three beams in unison are deflected both vertically and horizontally to produce the tri-colour raster.

On a monochrome (or colourless) transmission when only the $Y$ signal is present the three beams are modulated in the above-mentioned proportions corresponding to greys in the scene. This means that all the colours of a phosphor dot group, or triad, produce light, and the proportions are such that the light produced is white or, at least, close to the white that is given by an active monochrome tube screen. Grey shades are produced by reduced intensity white light.

On a colour transmission the $Y$ signal remains active to provide the basic scene in detail and brightness (i.e. luminance) while the $R$, $G$ and $B$ colour-difference signals vary in strength with respect to each other, and this alters the relative light intensity produced by each colour dot of a triad in accordance with the actual encoded colour information. If there is only fully saturated blue in the scene, for instance, the blue gun would be activated and only the blue dots on the screen would glow, and the *intensity* of the blue would be controlled by the level of the $Y$ signal. Similarly, for green and red scenes of full saturation.

In practice, very few colours are fully saturated; they are diluted a little or a great deal by the addition of white which, on the screen, is provided by illumination from *all* the dots in the proportions to give white light. The diluting *intensity* is, again, controlled by the $Y$ signal.

Detail in the picture displayed on the screen is thus given by the $Y$ signal, and the colour, which is boldly 'brushed in', so to speak, is given by the colouring signals. While the $Y$ or luminance signal operates over a bandwidth of some 5 MHz, the $R$, $G$ and $B$ colour-difference signals operate only within about 1 MHz bandwidth.

This lower colour definition is possible because human eyes are unable to define detail in colour. The technique, then, is to obtain the actual high definition in monochrome (luminance, that is, given by the $Y$ signal) and then to add the colour at much lower definition. This is also the reason why it is possible to use a screen composed of red-, green- and blue-glowing phosphor dots to give the colours. The eye is unable to distinguish between the separate colours in such small dimension as the dots; it integrates them to give white light, as we have seen, and the whole range of hues between red and blue, depending upon the relative intensities of the three colours. For instance, the eye perceives pink when the red dots are glowing and when this colour is diluted a little by white, given by the other two colours glowing just a little to desaturate the red. Pink, of course, is nothing more than desaturated red.

More recent colour display tubes incorporate phosphor stripes as distinct from phosphor dots, which glow red, green and blue. A special type of screening mask, as distinct from the perforated shadowmask, is used to resolve the colour display (see, for example, the author's *Colour Television Servicing*, second edition, Newnes–Butterworths).

Since the *colour-difference* signals are sometimes fed separately to the three grids of the tube, hence the terms red, green and blue grids, and the $Y$ signal to the three cathodes together, the signal-opposing actions between cathode and grid of each gun result in each electron beam being modulated only with its appropriate *primary-colour* signal, i.e. $R$, $G$ and $B$. The $Y$ part of each colour difference signal is thus effectively cancelled out. In this system the guns are said to act as 'colour matrices'; there is an alternative system whereby separate colour matrix stages are employed, each cathode then receiving three primary-colour signals and the grids providing the bias in the usual way.

**Colour bursts**

At the receiver, detection of the red and green colour-difference signals is possible only by adding the subcarrier to the detected sidebands. This is provided by a local reference generator, and it is imperative that this signal be phase-locked (and hence frequency-locked) to the suppressed subcarrier at the transmitter. Colour burst signals are generated and added to the television waveform on the back porch of the line sync pulses of the encoded colour signal at the transmitter, and these are directly locked to the phase – and frequency – of the subcarrier signal. These colour bursts, consisting of ten cycles of subcarrier frequency, are extracted after detection at the receiver and then channelled to a phase discriminator which also receives a sample signal from the local reference generator.

Should the phase of the reference generator signal differ from that of the burst signal, the discriminator gives a d.c. output which, after amplification, biases an electronic reactance (or capacitance diode) across the tuned oscillator of the generator, thereby locking the signal so that it has a phase relationship with the bursts; the discriminator output then falls to its nominal 'zero'. (Note: in any phase lock loop system of this kind the 'lock' occurs in phase quadrature; that is, the reference signal is 90 deg. out of phase with the sample signal.)

### Quadrature chrominance signal modulation

Quadrature modulation of the two chrominance (meaning colour) signals on the suppressed subcarrier at the transmitter can be represented by two carriers each amplitude-modulated, one by the red colour-difference and the other by the blue, but with one carrier displaced 90 deg. in phase relative to the other. The correct chrominance information at the set can only be obtained when the effective carriers are so displaced relative to the red and blue chrominance sidebands. This is why phase-locking is so important.

The chrominance modulation has components which vary both in amplitude and phase as the colour information changes. The amplitude components relate to *saturation*, while the phase components relate to *hue*. Colour sets are, for this reason, phase-sensitive; meaning that a change in phase of the signal in the equipment or during its passage from transmitter to receiver can adversely affect the colour display, changing it from its true colours.

### The PAL system

The American NTSC system is particularly phase-sensitive, but the refinements of the PAL system, developed by Dr. Walter Bruch of the German Telefunken Company, very successfully overcome the problem to a large extent by reversing the phase of the red chrominance signals on alternate lines, hence the term PAL, meaning Phase Alternate Line. Colour error on one line is cancelled out by a complementary colour error on the next line, the latter given by the swing of phase of the PAL system. The overall effect is that of correct hue display in spite of relatively high phase errors.

There are two basic techniques of achieving hue correction at the receiver. One is subjective, whereby the eye itself integrates the incorrect colour display and its complement, giving the impression of the correct colour display, and the other is electronic. A delay line in the receiver stores the colour information of one line of picture and releases it during the next line, along with the information of that line. The incorrect signal is then cancelled-out electronically by the stored signal of the previous phase-alternated line. The former system is called simple PAL or *PAL-S* and the latter delay-line PAL or *PAL-D*. The latter, of course, is by far the most satisfactory and is capable of compensating for quite large phase errors.

**Colour reception**

Fig. 8.2 gives a simple impression of the overall PAL colour system, from transmitter to receiver. The original *R, G* and *B* primary signals are 'seen' by the three appropriate guns of the tube because the *Y* signal is subtracted from the chrominance signal at each gun, as we have already observed.

It will now be apparent that the monochrome, colour and sound signals are all handled within a standard television channel. The sound is processed in exactly the same way as in a monochrome set, using the intercarrier technique, while the low-definition colour information is applied in the low-energy gaps that occur in the monochrome video spectrum centred on multiples of the line frequency. In other words, the chrominance signals are interleaved between high-energy bands of monochrome (or luminance) signal.

Fig. 8.3(*a*) shows the blue chrominance modulation on the subcarrier at the transmitter, while (*b*) shows the red chrominance modulation on a carrier that can be considered separate from that at (*a*) but displaced in phase by 90 deg. In practice, though, the one subcarrier serves both chrom-

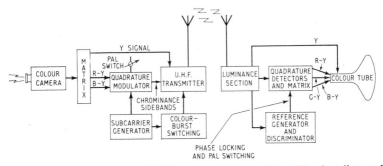

*Fig. 8.2.* Elementary colour television transmission system. The decoding end of the receiver section is shown in greater detail in Fig. 8.7.

# SERVICING WITH THE OSCILLOSCOPE

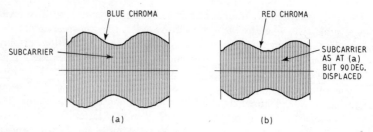

*Fig. 8.3.* Chrominance modulation envelopes: (*a*) blue and (*b*) red. The subcarriers upon which these signals are modulated have exactly the same frequency but are displaced in phase by 90 deg. to give quadrature modulation. This is achieved by adding the two modulations.

inance signals and a 90 deg. phasing network is interposed relative to each section of the quadrature modulator.

Fig. 8.4(*a*) shows how the chrominance signal is superimposed, so to speak, upon the luminance information of the video waveform, while (*b*) shows the chrominance sidebands centred upon the subcarrier frequency of 4.43361875 MHz, with the actual subcarrier suppressed.

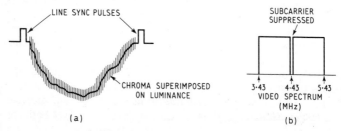

*Fig. 8.4.* (*a*) Showing how the chrominance signal is superimposed upon the luminance information of the composite video signal. (*b*) The chrominance upper and lower sidebands centred on the subcarrier frequency (the subcarrier being suppressed at the transmitter).

Fig. 8.5(*a*) shows one line of colour-encoded signal and the colour bursts on the back porches to the line sync pulses, while (*b*) gives waveforms of the field sync and colour bursts at the beginning of even and odd fields. These, of course, relate to the British 625 standard.

### Scope testing in colour sets

Scope testing in the video sections of colour sets follows exactly the techniques expounded in previous chapters of this book. The overall *Y* (lumin-

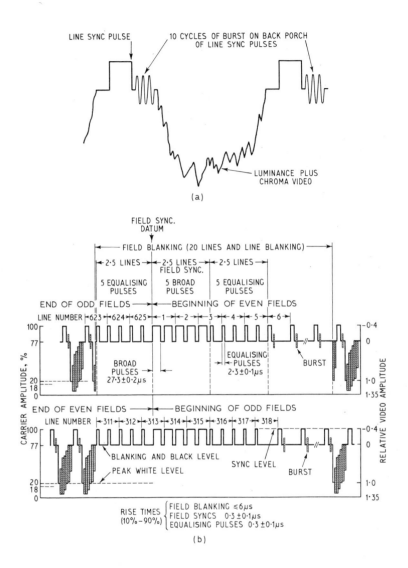

*Fig. 8.5.* At (*a*) are shown the colour bursts on the back porch to the line sync pulses. Sync signals at the beginning of even and odd fields are shown at (*b*) – the latter by courtesy of the BBC. Ceefax information is transmitted on blanked lines 17 (330) and 18 (331) at a bit rate of 7 megabits per second (see Chapter 3).

ance) or monochrome signal at video can be seen at the cathode of the colour tube, while the separate colour-difference signals can be monitored at the appropriate colour grids. When there is no colour in a transmission, the chrominance signals fall to zero, and d.c. restoration is used in the sets to hold the grids at a specific potential relative to the $Y$ signal at the cathodes. D.C. restoration is more important in colour sets than in contemporary monochrome counterparts to ensure accurate colour representation and grey-scale tracking over swings of video signal.

A scope is also extremely useful for checking the reference oscillator at the receiver, and for ensuring that this is locked by the action of the phase discriminator and electronic reactance (usually a capacitance diode, the capacitance value of which decreases with increase in reverse bias applied to it – that is, as the depletion layer in it is enlarged).

Square-wave testing can be adopted in the colour-difference amplifiers, as well as in the luminance amplifier, in accordance with the techniques already described. While the luminance amplifier must have a wide passband to handle the high-definition luminance signals, the colour-difference amplifiers need only pass a little above 1 MHz, and for this reason their h.f. square-wave performance may not be so exacting. However, the phase characteristics need to be carefully matched, for differences can result in horizontal displacement of the three colours making up a picture element, and this applies also to receivers where *RGB* (primary colour) drive is used and where a separate primary colour amplifier is used to carry signals of corresponding colour to each of the display tube modulation electrodes (grid or cathode, the latter being favoured since this gives an increase in sensitivity of some 30 per cent over grid drive; that is, the dynamic '$g_m$' is some 30 per cent greater).

The scope can be used for tracing the video signals from the luminance detector into the various luminance and colour-difference channels, and it can also be used to trace the burst signals from the output of the burst filter. Indeed, with the present trend in colour receiver design focused towards circuit modules, the scope assumes a new importance in colour set fault diagnosis, since it makes it relatively easy to check the sound and video (both monochrome and chrominance) from module to module without having to disconnect or disturb the faulty set too much.

Service technicians need a dot and crosshatch generator for setting up the convergence controls (see later), and some such generators also provide colour bars. The signals applied to the set with the generator switched to 'colour bars' (or equivalent) are extremely useful for signal tracing with the scope.

For checking the monochrome performance of a colour set, any monochrome generator or off-the-air signal is suitable, and the tests follow along the lines of those described in past chapters for monochrome-set testing.

## PAL decoder waveforms

The waveforms in Fig. 8.6 show how signals from a colour-bar test transmission appear on an oscilloscope with a $Y$ bandwidth of about 6 MHz at various points in the receiver, and these are related to the various stages of a PAL decoder block diagram depicted in Fig. 8.7.

The waveform at (a) is at the luminance detector output, and this is applied to the chroma bandpass filter deleting most of the luminance signal, leaving

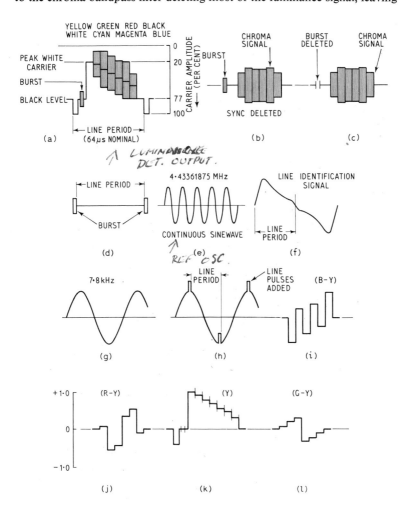

*Fig. 8.6.* Waveforms in the various sections of a PAL decoder (see Fig. 8.7).

141

## SERVICING WITH THE OSCILLOSCOPE

chroma only as waveform (*b*). Some chroma is fed to the burst take-off detector circuit, which passes burst signals only (*d*), while the bursts are deleted from the chroma signal at (*c*). The reference oscillator signal waveform is given at (*e*). Also derived from the burst detector circuit is waveform (*f*) which is used to identify the phase of the $R-Y$ signal which is reversed on alternate lines at the transmitter as already explained. This signal shock-excites the 7·8 kHz (half-line frequency) high-$Q$ tuned circuit, giving the sine wave at (*g*). Added to this signal are pulses from the line timebase and the combined waveform (*h*) is used to 'trigger' the PAL alternating inverter stage in synchronism with the PAL switching at the transmitter.

The blue and red colour-difference colour-bar test waveforms are shown respectively at (*i*) and (*j*). Waveform (*k*) is essentially that of the luminance signal subsequent to the subcarrier notch filter which is incorporated in the luminance channel (not shown in Fig. 8.7), while that at (*l*) is the green chrominance signal.

Although these waveforms are not real c.r.t. traces they do, nevertheless, give a close approximation to the displays that are obtained at the various

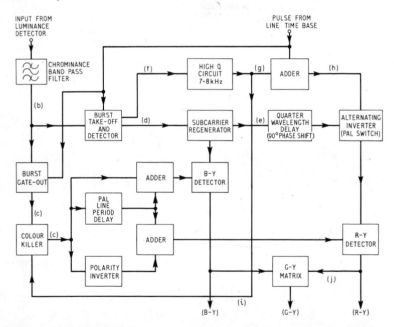

*Fig. 8.7.* Block diagram of a PAL decoder. The lower-case letters correspond to the waveforms in Fig. 8.6. The green chrominance signal is obtained by matrixing in correct proportions the blue and red chrominance signals. (Figs. 8.6 and 8.7 are based on Thorn-A.E.I. information).

points in the decoder part of a PAL-D receiver when the encoded input is from a colour-bar generator having signal characteristics similar to those of waveform (*a*). Off-oscilloscope-screen displays are given later in this chapter.

It should be noted that the colour-killer is effectively a 'gate' circuit which opens only when a colour transmission is present (i.e. a signal containing colour bursts). It closes on monochrome transmissions, thereby preventing the chrominance circuits from giving spurious outputs and 'colouring' a black-and-white picture.

While the PAL alternating inverter is effectively triggered by the pulses fed in from the line timebase (waveform *h*), the actual lines which are phase inverted need to be matched to those same lines at the transmitter. This is where the identification ('ident') signal from the colour bursts – which themselves are alternated at the transmitter – takes over. This signal does not actually synchronise the line phase-switching, but it makes sure that the phase-changed lines at the receiver correspond to those same lines at the transmitter. The signal 'programmes', so to speak, the mark-space *timing* of the trigger alternating switch, by the phase-swings of the colour bursts, to the line timebase switching pulses.

### Colour receiver timebases

Colour sets have line and field timebases which, apart from power, are virtually identical to those in monochrome sets. Dual-standard colour sets also have line timebase switching (10,125 Hz for 405 lines and 15,625 Hz for 625 lines), but owing to the greater effective 'stiffness' of the three electron beams of the colour picture tube, given by an e.h.t. of about 25 kV, greater power is needed to deflect them fully over the screen. The line timebase also has to supply the 25 kV of e.h.t. for the three beams, rising to a maximum of about 1·5 mA. The timebases thus have to work at peak efficiency, and it is also essential that their scan linearity is adjusted to a high order to ensure that the three beams remain in registration with the appropriate colour phosphor dots over the whole area of screen.

Faults in the timebases can be traced with the scope, and drawings of oscillograms to be expected in various parts of the circuits are now given in the majority of colour-set service manuals. These are identified in terms of both amplitude and sweep velocity.

### Purity and convergence

The timebases of colour sets also have to provide extra signals for dynamic convergence, which are not required by monochrome sets. It was explained earlier that a colour tube incorporates three electron guns, producing three electron beams, one for each colour phosphor. To ensure that the beams impinge only upon phosphor dots corresponding to their own colour,

## SERVICING WITH THE OSCILLOSCOPE

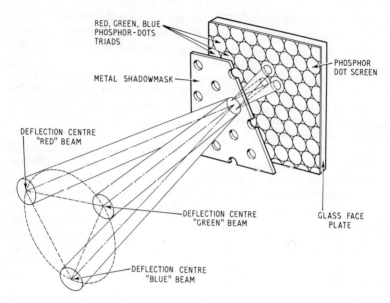

*Fig. 8.8.* Showing the principle of the shadowmask in a colour tube (see text).

a mask carrying a great many small holes (one for each group of three phosphor dots) is fitted inside the tube, a little distance behind the phosphor-dot screen, and arranged geometrically so that when the three beams are corrected by tube-neck magnetic fields for 'approach angle' and deflection centre they will pass through the holes and strike only the phosphors of matching colour-glow. The idea is illustrated in Fig. 8.8.

The perforated mask is generally known as a *shadowmask*, since if light beams instead of electron beams were present the mask would produce shadows on the phosphor-dot screen at all points other than those of the dot-triads corresponding to the beam colours. The colour picture tube is therefore called a shadowmask tube, and its component parts are shown in Fig. 8.9.

A magnetic field is caused to cut the three beams *together* within the tube neck, and its intensity can be adjusted (by the use of two ring magnets whose fields are added or opposed when one ring is rotated relative to the other) to apply a fixed deflection to the three beams, ensuring that they approach the mask at the correct angle required to pass through the holes and activate only the phosphor dots of the corresponding colours. This is called a *purity magnet*.

Each beam *separately* is also subjected to a fixed magnetic field. Thus,

144

## COLOUR TELEVISION WAVEFORMS

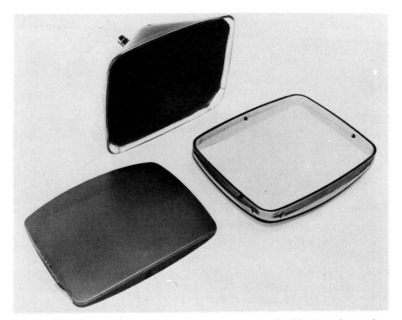

*Fig. 8.9.* Shadowmask, faceplate, and cone and neck of a Mazda colour tube.

there are three more magnets, one for each beam, and these are separately adjustable to shift the beams radially within the neck so that they together pass through mask holes to activate adjacent groups of triads (about three in practice) when the beams are deflected within a small centre region of the screen. These are called *static convergence magnets*, and a further one, called a *blue lateral magnet*, allows the blue beam only to be shifted laterally so that the blue spot of light can be shifted slightly in a horizontal plane on the screen, making it possible to get all three – red, green and blue – spots of light in correct registration. This is, in fact, the exercise of *static convergence adjustment*. That is, getting all the spots coincident in the screen centre, to produce white spots when the inputs are in the correct proportions.

**Dynamic convergence**

A problem occurs when the three beams are deflected horizontally and vertically to produce the raster, for then the beam landings tend to fall away from the correct colour phosphors towards the edges of the screen. The reason for this is twofold. One, because the screen is relatively flat compared with the radius of beam deflection, and two because the three guns cannot be placed on a common axis at the end of the tube neck (see Fig. 8.10).

The problem is solved by the use of electromagnets whose changing

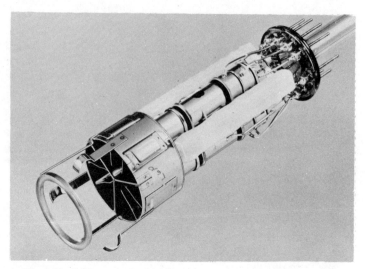

*Fig. 8.10.* Triple-gun of a Mazda colour tube. This illustration shows clearly how the guns are displaced from the common tube axis.

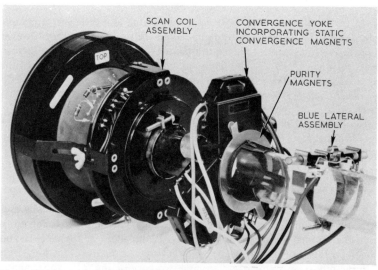

*Fig. 8.11.* The neck items on a Mazda colour tube. More recent colour tubes are 'self-converging' by the use of a precision convergence yoke which is treated as an integral part of the tube system.

fields are concentrated at the same point as those emanating from the permanent static convergence magnets. The electromagnets are called *dynamic convergence coils* or yokes, and they are energised by corrective currents derived from the line and field timebases. The neck assembly of a colour tube, including scanning coils, purity and convergence units, is shown in Fig. 8.11.

Because zero dynamic correction is required at the screen centre (this has already been handled by the static magnets) the dynamic convergence coils are fed with a current waveform of parabolic character, as shown in Fig. 8.12(*a*). This ensures that maximum correction occurs at one deflection extreme, falling to zero at the centre and then rising to maximum correction again at the other deflection extreme.

The effect produced by the three guns being displaced from the tube axis is countered by adding a little sawtooth waveform to the parabolic current, Fig. 8.12.(*b*) Fig. 8.12(*c*) shows the compounded parabolic and sawtooth waveforms when the sawtooth phase is positive and (*d*) when the phase is negative. The effect is that the parabola is tilted a little one way by the positive sawtooth and a little the other way by the negative sawtooth. The degree of tilt depends, of course, on the amplitude of the sawtooth relative to the amplitude of the parabolic wave.

The dynamic convergence control panel of colour receivers contains twelve or more preset adjustments to set the relative amplitudes and signal phases for optimum convergence over the entire screen area (but recent tubes, with integral precision yokes, are significantly reducing the number). These adjustments are rather critical and interdependent, and are best carried out in accordance with the instructions supplied by the particular manufacturer. A generator giving, at least, a selection of cross-hatch patterns is essential for dynamic convergence adjustments; but some degree of adjustment is possible by using the colour Test Card F.

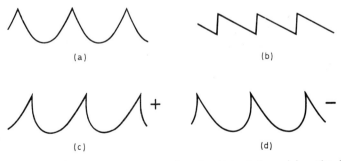

*Fig. 8.12.* Dynamic convergence correction is attained by mixing the basic parabolic waveform (*a*) with a sawtooth waveform (*b*) in positive or negative phase to produce waves (*c*) and (*d*) respectively.

# SERVICING WITH THE OSCILLOSCOPE

It falls outside the scope of this chapter to discuss the various circuits employed for dynamic convergence and adjustment, but the prime considerations are that the basic components required to produce the non-linear correcting waveforms are a linear sawtooth one and a non-linear parabolic one, at both line and field frequencies, and that these in adjustable degrees of amplitude and phase need to be passed through the convergence coils.

It is possible to obtain the signals by applying the required *voltage* across the coils, in which case the convergence coil and scanning coil are effectively in parallel. This scheme demands high-inductance convergence coils to minimise the loading on the timebase. Another method arranges for the basically linear scanning *current* (that in the scanning coils) to pass through the convergence coils, by the coils being series-connected. This scheme calls for low-inductance convergence coils to minimise power loss. Both schemes have been used in practice, but the net result is always the creation of essentially parabolic correction currents which can be displayed on the scope screen by the methods already referred to in this book. For instance, the current waveform in the convergence coils can be monitored in terms of voltage developed across a low value resistor connected in series with them.

A field parabolic waveform is present at the cathode of the field output valve or transistor emitter, and it is from this source that the field convergence signal is often obtained. In the line circuit the voltage across an inductor is sometimes used.

Waveforms in these circuits also tell whether the convergence controls are working correctly, for they should alter both in amplitude and phase (giving tilt), as already explained, as the various dynamic convergence controls are adjusted.

**Pincushion correction**

Correction for pincushion distortion on the screen of monochrome sets is achieved by the use of small bar magnets located round the tube flare, often on the scanning coil assembly. These are simply orientated so that their fields pull out the concaves at the picture edges. This technique cannot be employed in colour sets because any spurious magnetic field in the vicinity of the shadowmask tube will impair both purity and convergence.

The problem can be solved by the use of a special *transductor unit*. This, which is rather a special kind of transformer, is coupled into the line and field output circuits, and the currents so derived are shaped and then fed back again to the field and line scanning coils respectively. Line sawtooth current is applied to a pair of windings on the transductor which are connected series-antiphase, while the field sawtooth current is applied to a third winding between them (Fig. 8.13).

When field current flows through the third, middle winding, the satura-

## COLOUR TELEVISION WAVEFORMS

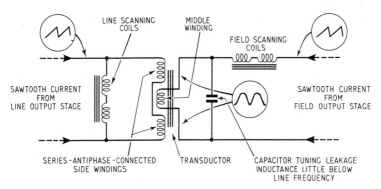

*Fig. 8.13.* Showing the basic connections of a transductor pincushion distortion correcting device and the waveforms involved.

tion of the core changes at field frequency, and this causes a synchronised change in inductance of the two side windings. Now, because the side windings are connected in parallel with the line scanning coils, a proportion of the energising current is bypassed through them from the line output stage, and the proportion bypassed changes with field scan, as the transductor core saturation changes. Thus, the line scan amplitude is affected over the field scan in such a manner that the curves at the sides of the picture or raster are straightened.

Correction at the top and bottom horizontal edges is achieved by the transductor translating the sawtooth current in its two side windings into an S-shaped voltage wave across its third or middle winding. Normally, this would cause a parabolic current wave at line frequency to flow in the field scanning coils, along with the ordinary sawtooth scanning current, but of a polarity that would emphasize the pincushion effect. The polarity is reversed, to cause pincushion correction, by tuning the transductor's middle-winding leakage inductance to resonate a little below line frequency. This produces a correctly-phased sine wave sufficiently close to a parabolic waveform to give the required correction.

Fig. 8.13 shows the basic connections to the transductor; also the capacitor tuning the middle winding leakage inductance towards the line frequency. On dual-standard sets, a second capacitor is switched in to change the tuned frequency to correspond to the selected line frequency. Basic waveforms which can be expected in this kind of circuit are also indicated. More recent colour receivers employ wide-scanning-angle picture tubes, which call for more elaborate dynamic convergence circuits to minimise red, green and blue raster displacements at the scan extremes; additional circuits may also be used to minimise the exaggerated pin-cushion distortion which also tends to result from the wider angle of deflection. These circuits

## SERVICING WITH THE OSCILLOSCOPE

and the new tubes are dealt with in some detail in the later volumes of the author's *Newnes Colour TV Servicing Manuals* (Newnes-Butterworths), while a full description of the colour television system, with an emphasis towards aspects of servicing, is given in the author's *Colour Television Servicing*, second edition (Newnes-Butterworths).

We have not attempted here to describe in detail how colour receivers work. the aim has been to discuss the actual waveforms that are encountered in

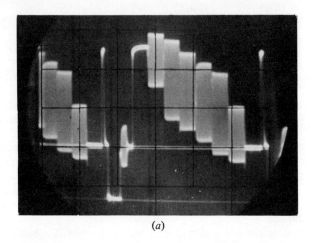

(a)

*Fig. 8.14.* (*a to h*) A series of colour signal waveforms as obtained direct from the screen of an oscilloscope. These are fully described in the text.

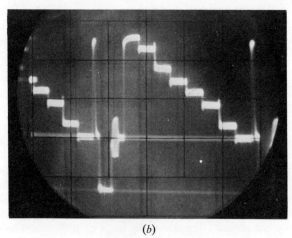

(b)

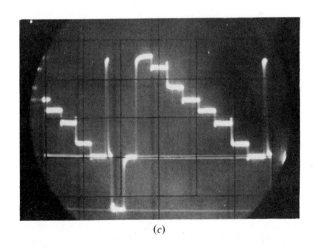

(c)

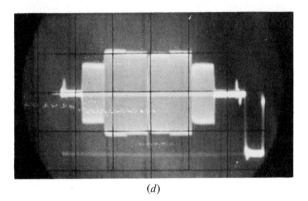

(d)

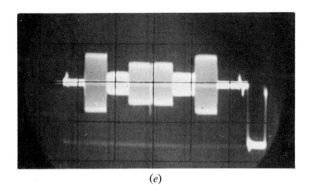

(e)

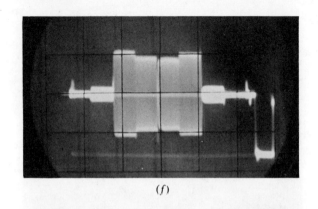

(f)

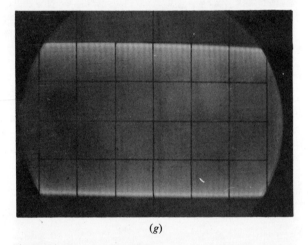

(g)

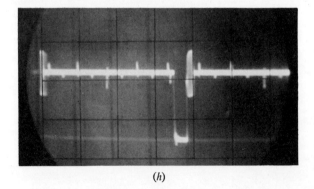

(h)

colour sets, thereby assisting technicians in tracing faults in this kind of set, aided, of course, by the oscilloscope.

As a further aid in this connection, in Fig. 8.14 is given a series of colour television waveforms which were actually photographed from the screen of an oscilloscope during a servicing operation on a colour set, while the set was being fed with signal from a standard colour bar generator.

At (*a*) is shown the composite and chroma signals – with the burst – over $1\frac{1}{2}$ lines. At (*b*) is shown the signal with the chroma switched off but with the bursts still present, while (*c*) is signal with the bursts switched off, too. Oscillograms (*b*) and (*c*) are thus representative of the luminance signal, due to the colour bars only. These signals would be picked up from the output of the vision detector in the majority of sets.

The $(R-Y)$ and $(B-Y)$ chroma signal is shown in (*d*), while the $(R-Y)$ and $(B-Y)$ chroma signals are shown separately in (*e*) and (*f*) respectively. These signals would be picked up at the output of the chroma bandpass amplifier, the latter two with the $(B-Y)$ and $(R-Y)$ modulation at the generator switched off alternately. These sorts of waveform tell how well the PAL decoder is adjusted. Waveform (*g*) is of the reference generator, while (*h*) shows the burst of one line with the video signal deleted.

**Other colour television oscillograms**

The series of oscillograms in Fig. 8.15 refers to a recent receiver (BRC 9000 series chassis, which is of fully modular construction and comprises three large printed circuit boards plus five smaller ones; the chassis is also, with the exception of the display tube, all solidstate, using transistors and integrated circuits) when it is in receipt of a colour bar signal and correctly adjusted to display this signal. The oscillograms were taken with a 10 M/11 pF 10:1 attenuation ratio low-capacitance probe correctly compensated, and the vertical and horizontal scale values given in the captions are the actual values, i.e. with the attenuation of the probe taken into account. Each waveform is detailed in its own caption. See also Fig. 5.37.

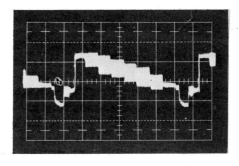

*Fig. 8.15(a).* Detected video signal. Amplitude 1 V/div, sweep 10 μs/div.

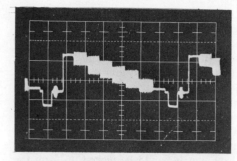

*Fig. 8.15(b).* Video input to luminance integrated circuit. Amplitude 1 V/div, sweep 10 μs/div.

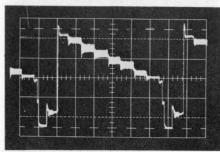

*Fig. 8.15(c).* Luminance delay line drive with chroma mostly deleted. Amplitude 2 V/div, sweep 10 μs/div.

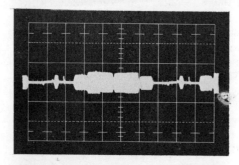

*Fig. 8.15(d).* Output from chroma filter with Y signal removed, showing bursts. Amplitude 0·5 V/div, sweep 10 μs/div.

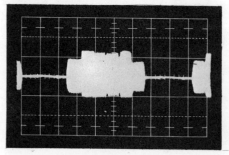

*Fig. 8.15(e).* Chroma delay line drive after amplification by chroma bandpass amplifier, with bursts blanked. Amplitude 0·5 V/div, sweep 10 μs/div.

*Fig. 8.15(f). B−Y* output. Amplitude 100 mV/div, sweep 10 μs/div.

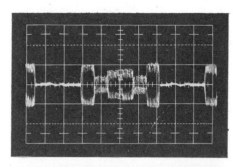

*Fig. 8.15(g). R−Y* output. Amplitude and sweep as (*f*).

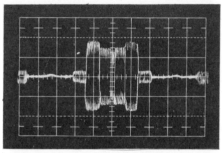

*Fig. 8.15(h).* 4·4 MHz reference signal. Amplitude 200 mV/div, sweep 10 μs/div.

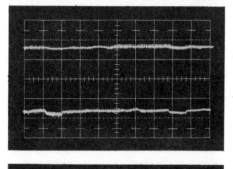

*Fig. 8.15(i).* 7·8 kHz half line frequency square-wave signal. Amplitude 0·5 V/div, sweep 20 μs/div.

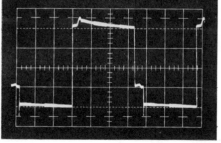

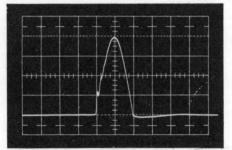

*Fig. 8.15(j).* Shaped retrace pulse (line). Amplitude 2 V/div, sweep 5 µs/div.

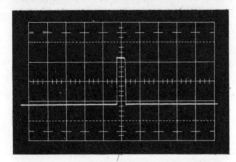

*Fig. 8.15(k).* Composite sync signal with scope sweep adjusted to show line sync pulse. Amplitude 10 V/div, sweep 10 µs/div.

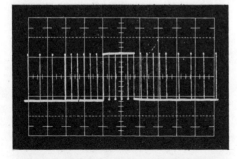

*Fig. 8.15(l).* Composite sync signal with scope sweep adjusted to show field sync. Amplitude 10 V/div, sweep 100 µs/div.

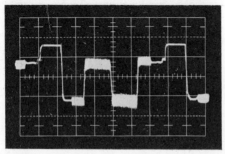

*Fig. 8.15(m).* Red video (primary colour; the receiver employs *RGB* drive to the display tube) output. Amplitude 50 V/div, sweep 10 µs/div.

*Fig. 8.15(n).* Green video output. Amplitude and sweep as (*m*).

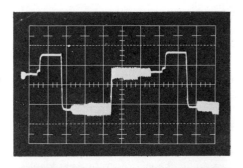

*Fig. 8.15(o).* Blue video output. Amplitude and sweep as (*m*).

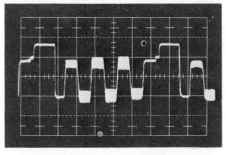

*Fig. 8.15(p).* Line output transistor collector waveform. Amplitude 200 V/div, sweep 10 μs/div.

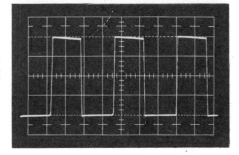

# 9: STEREO RADIO WAVEFORMS

It is well known that stereophonic sound reproduction demands at least two audio channels operating in complete isolation from the sound source to the loudspeakers. Domestic stereo systems use two channels, called the left-hand and right-hand channels or $A$ and $B$ channels respectively.

**Stereo principles**

The signals for these two channels are obtained from two microphones at the source. The microphones may be in a common housing, forming the so-called stereo microphone, with their axes of maximum response at right-angles to each other, or they may be two separate mono units of ordinary type, orientated so that one picks up the sounds emanating from the left-hand side of the source while the other deals with the sounds on the right-hand side.

The signals from the left- and right-hand sides of the source are thus fed into the $A$ and $B$ channels and they are ultimately reproduced at two loudspeakers, one placed to the left of the listener working from the signals in the $A$ channel, and the other placed to the right working from the signals in the $B$ channel. How well the stereo effect is produced depends upon the original orientation of the two microphones or elements of a stereo microphone, on the degree of isolation between the $A$ and $B$ channels and on the placement of the loudspeakers in relation to the listener. It is not here intended to delve into these problems, and readers interested in the wider aspects of f.m. stereo reception and reproduction may find my books *F.M. Radio Servicing Handbook*, second edition, and *The Audio Handbook* of interest; both are published by Newnes–Butterworths.

When stereo is recorded, the two isolated channels are maintained at the recording medium. On disc, the single groove carries both the $A$ and $B$ information, while on tape two tracks are used. The former requires a special stereo pick-up cartridge and the latter two isolated replay-head sections.

**Stereo broadcasting**

Stereo via radio also demands two isolated channels, and since it is neither feasible nor good economics to tie up two separate transmitters for one

# STEREO RADIO WAVEFORMS

stereo programme the radio system is based on a single-transmitter multiplexing technique, where both the $A$ and $B$ signals are effectively carried in isolation on one v.h.f. carrier. Like colour television, the scheme has to be compatible. This means that the stereo-encoded transmission must give correct mono balance on ordinary non-stereo receivers and tuners.

It will be appreciated that mono balance of the stereo signals is obtained when the signals in the left and right channels are combined. Thus we have mono equals $A + B$. By the same token, it follows that the stereo information must be the difference between the signals in the left and right channels. Thus stereo equals $A - B$. There is no stereo information when the $A$ and $B$ signals are the same, and greater stereo information as the difference between the two signals increases.

Straight away, therefore, the mono information can be fed to a v.h.f.-f.m. transmitter in the form $A + B$, and any mono sets picking up this signal will work normally. The problem is to add the $A - B$ information, and this is where the multiplexing technique comes in. The $A - B$ information is amplitude-modulated on to a 38 kHz subcarrier at the transmitter and, like colour television (see Chapter 8) the subcarrier itself is suppressed, leaving only the $A - B$ sidebands; these are fed to the f.m. transmitter along with the $A + B$ mono modulation.

Since demodulation of the $A - B$ sidebands at the receiver requires the original subcarrier, this is recreated by the stereo decoder end of the receiver producing its own 38 kHz signal, which can then be added to the f.m. detected $A - B$ sidebands. This results in retrieval of the original $A - B$ audio information.

Before we see how this is processed in conjunction with the $A + B$ information to give the $A$ and $B$ signals in isolation, it should be understood that the recreated 38 kHz subcarrier must be synchronised to the subcarrier originally used at the transmitter. This is achieved by the transmitter sending out a synchronising signal, called a *pilot tone*. The frequency of this is half that of the subcarrier – that is 19 kHz – and the decoder at the receiver either doubles it to produce the required 38 kHz subcarrier or uses it to 'lock' or synchronise a 38 kHz reference generator. In both cases the missing 38 kHz subcarrier is produced in the decoder and synchronised with the original subcarrier at the transmitter.

### Transmitter waveforms

Fig. 9.1 shows the waveforms involved in the transmitting process. The audio signal that may be present in the $A$ channel is shown at (*a*), the $B$ channel signal at (*b*), the $A + B$ signal at (*c*), the modulated $A - B$ signal at (*d*) and the pilot tone signal at (*e*). All these signals frequency-modulate the v.h.f. transmitter, and the resulting modulation spectrum is given in Fig. 9.2.

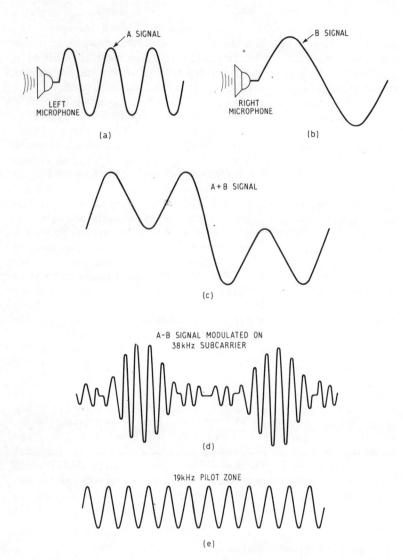

*Fig. 9.1.* Signals at the transmitter end of a radio stereo system (see text).

# STEREO RADIO WAVEFORMS

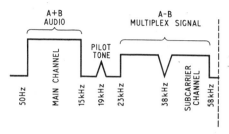

*Fig. 9.2.* The signals within this spectrum frequency-modulate a v.h.f. transmitter for stereo. These same signals appear across the detector output of an f.m. tuner or receiver when the de-emphasis network is disconnected.

A v.h.f.-f.m. receiver or tuner, provided the f.m. detector is suitable and the i.f. stages accurately aligned with sufficient pass band (see later) will deliver the $A + B$ information at its output and also the $A - B$ sidebands and pilot carrier signal when the ordinary f.m. de-emphasis network is disconnected. Normally, the de-emphasis network attenuates demodulated signal components other than the $A + B$ signals, so when a stereo decoder is used with a suitable mono f.m. set or tuner the de-emphasis must be disconnected and reintroduced at the output of each $A$ and $B$ channel.

It might well be wondered why the original 38 kHz subcarrier is suppressed at the transmitter while a pilot carrier signal is added. The reason is that a full-amplitude subcarrier would restrict the $A + B$ and $A - B$ modulation depth, because the subcarrier itself would account for a substantial deviation. The pilot tone signal need only be of very small amplitude, thus taking up just a little deviation percentage, since it can be amplified easily (and frequency-doubled) at the decoder.

By the same token, it may be thought that the $A - B$ sideband modulation on the main v.h.f. carrier would restrict the modulation deviation available for the $A + B$ signals. This does not happen in practice because the $A + B$ modulation generally rises when the $A - B$ modulation falls, and vice versa.

The effective modulation depth taken by the pilot carrier is about one-tenth that of the full $A + B$ and $A - B$ modulation (20 dB down), thereby allowing 90 per cent for the full mono plus stereo modulation. Even so, the signal-noise performance on a stereo transmission is some 20 dB poorer than a mono transmission under threshold signal conditions, and for this reason the aerial system assumes a greater importance for v.h.f.-f.m. stereo reception than for similar mono reception. When a *mono* receiver or tuner is working from a stereo-encoded signal, the signal-noise performance is worsened relative to a pure mono transmission only by about 4 dB.

Fig. 9.3 gives a block diagram of a typical f.m. tuner with a stereo decoder

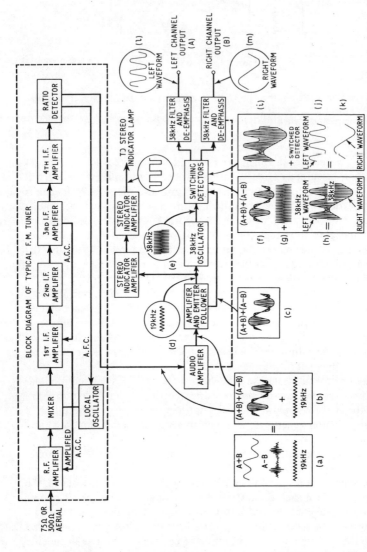

*Fig. 9.3.* Block diagram of f.m. tuner and connected stereo decoder of the *time multiplex* type, showing waveforms. The procedure nowadays is to use a 38 kHz doubler stage, which doubles the 19 kHz pilot tone signal, rather than a synchronised 38 kHz oscillator.

connected to it, and the waveforms involved. The decoder starts off with an audio amplifier which lifts the multiplex signals, and the output of this is connected to an amplifier and emitter-follower stage. This delivers an amplified pilot signal to lock the 38 kHz oscillator, and the multiplex signals less the pilot carrier.

The multiplex signals along with the oscillator output are applied to the switching detectors, from whence the original $A$ and $B$ signals are obtained and then fed to the stereo amplifier system through de-emphasis filters.

Most stereo decoders have a stereo indicator lamp, which lights when the tuner is responding to a stereo transmission. This lamp in the arrangement shown is switched on by a transistor circuit which is activated by the 19 kHz pilot carrier. Thus as soon as a transmission containing a pilot carrier is received (that is, a stereo transmission) the transistor circuit switches on and the lamp lights. The block diagram shows this system.

This is called the *time multiplex decoding system*, and the 38 kHz oscillator shown in Fig. 9.3 is now commonly a frequency doubler which, by the use of diodes or a transistor, doubles the 19 kHz pilot tone signal to yield the 38 kHz subcarrier that is used for switching the synchronous detectors. More recent time multiplex decoders are based on the phase lock loop technique, and integrated circuits are now readily available that integrate the functions shown in the 'blocks' of Fig. 9.4(a). Here the loop consists of a phase-locked comparator, low-pass filter, d.c. amplifier, 76 kHz voltage-controlled oscillator (v.c.o.) and two divide-by-two stages.

The output from the v.c.o. is frequency-divided twice to yield a 19 kHz signal, and this is compared with the pilot tone component of the composite input signal (see Fig. 9.2) by the phase comparator, whose output is fed back to the v.c.o. (the frequency of the v.c.o. is adjustable by a direct voltage) which maintains a phase-locked condition in the presence of a stereo broadcast. As in all phase-locked loop systems (including the reference signal phasing of colour receivers), the 'locked' condition obtains when there is a 90 degree phase difference between the loop and the input signal.

The pilot tone presence-detector multiplies (i.e. combines) the composite signal with the 19 kHz signal from the divide-by-two stage (preceded by a phase corrector that adjusts the 19 kHz signal to the pilot tone phase) to yield a direct voltage. The direct voltage component is extracted by the low-pass filter, and is then used to operate a Schmitt trigger. This in turn operates the stereo on/off gate and the stereo indicator lamp control circuit. When a direct voltage component is not present at the trigger (i.e. on a mono transmission), the stereo on/off gate disables the decoding stage and mono signals are then passed to the left and right audio output channels.

An alternative method of stereo decoding is by the so-called *frequency multiplex system*, the block diagram of which (as used in the Mullard LP1400 Stereo Decoder Module) is given in Fig. 9.4(b).

## SERVICING WITH THE OSCILLOSCOPE

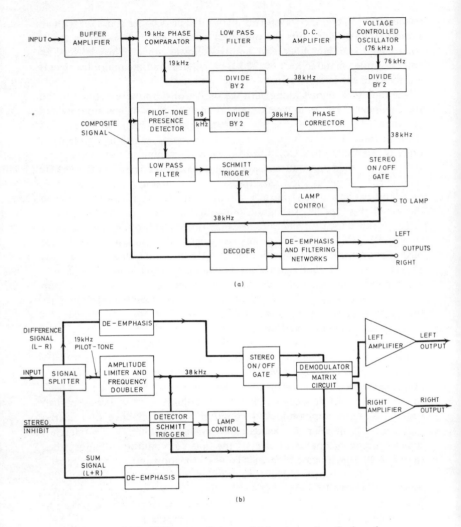

*Fig. 9.4.* (*a*) Block diagram of *time multiplex* stereo decoder based on a phase-locked loop integrated circuit (see text). (*b*) Block diagram of *frequency multiplex* stereo decoder (after Mullard Ltd).

The majority of decoders at the time of writing, however, are based in one way or another on the time multiplex system, though there are advantages in the frequency multiplex system (see, for example, H. Q. N. Davies, *Mullard Technical Communications*, Volume 13, No. 122, April 1974).

## STEREO RADIO WAVEFORMS

**Decoder waveforms**

Now let us study the waveforms of Fig. 9.3. All these can be displayed on the screen of a scope suitable for audio applications when the tuner-decoder set-up is receiving a suitable multiplex signal. Multiplex generators are available for facilitating waveform checks of this nature and for aligning decoder circuits; but test waveforms are also transmitted at certain times by the BBC for stereo receiver and tuner channel-to-channel testing. Test tones are sometimes transmitted separately in the $A$ and $B$ channels and in the $A$ and $B$ channels together. The former tests make it possible to appraise the channel isolation of the decoder. For instance, signals in the $A$ channel should not break through significantly to the $B$ channel, and vice versa. The scope can be used to monitor these test signals, and to measure the ratio of strengths of the $A$ signal in the $A$ channel to that same signal in the $B$ channel, and vice versa. This gives the channel isolation performance which, for good stereo reproduction, must be at least 20 dB over the audio spectrum and 30–40 dB at mid-spectrum.

The waveforms at (*a*) are the three signal components across the f.m. detector load of the tuner or receiver, namely, the $A + B$ mono component, the $A - B$ sidebands amplitude-modulated on the 38 kHz subcarrier, with the subcarrier suppressed and the 19 kHz pilot tone signal. The integration of the $A + B$ and the $A - B$ components is shown at (*b*) along with the pilot carrier.

Waveform (*c*) shows the integrated $A + B$ and $A - B$ waveforms less the pilot carrier at an output of the emitter-follower, while (*d*) shows the isolated pilot carrier. The 38 kHz signal waveform is shown at (*e*), locked by waveform (*d*). Waveform (*f*) is again the integrated $A + B$ and $A - B$ signals and this appears at the switching detectors along with the 38 kHz recreated subcarrier. The integration of waveforms (*f*) and (*g*) gives the complex waveform (*h*), and it will be seen that one modulation envelope of this waveform represents the audio signal in the $A$ channel and the other envelope the audio signal in the $B$ channel.

The complex signal, shown again at (*i*), is detected alternately on its positive and negative envelopes by the detectors being switched at a rate of 38 kHz by the pilot carrier signal. Thus on one switching cycle the left audio signal appears in the $A$ output (waveform *j*), and on the other switching cycle the right audio signal appears in the $B$ output (waveform *k*). These waveforms are shown again after the de-emphasis filters at (*l*) and (*m*) respectively.

We have thus seen that the $A$ and $B$ signals that originally modulated the stereo encoder at the transmitter (Fig. 9.1) appear in similar form, still in isolation, at the left and right outputs of the stereo decoder at the receiver. The stereo requirements have thus been met.

The mathematics of the system are quite straightforward, and for those

## SERVICING WITH THE OSCILLOSCOPE

readers who may be interested in this respect $(A + B) + (A - B) = 2A$ and $(A + B) - (A - B) = 2B$, from whence the original $A$ and $B$ signals are obtained by the effective plus and minus switching of the detectors.

### Importance of correct alignment

It should be understood that the decoding system in any form will only operate properly with maximum inter-channel separation and least noise and distortion when the locally produced 38 kHz subcarrier is accurately phased to the suppressed subcarrier at the transmitter. Thus the tuning of the reference oscillator is important. The pilot tone, of course, phase-locks the reference generator signal as well as synchronising its frequency, but while the frequency may be synchronised the phase could still be out a little if the tuning of the various coils and transformers in the decoder is in error.

Some off-the-screen oscillograms are given in Fig. 9.5: (*a*) shows full-carrier stereo modulation when $A$ equals $B$; (*b*) shows full modulation of the $A$ signal only; (*c*) full modulation of $A - B$; and (*d*) the effect on (*b*) of poor phase characteristics in the tuner's i.f. channel. Fig. 9.6(*c*) shows another off-screen display of stereo multiplex, this time with rather bad distortion on the audio modulation (400 Hz tone) and poor separation characteristics.

### Bandwidth

For accurate stereo decoding coupled with the least harmonic distortion at high deviation (modulation) levels the i.f. bandwidth should be around 240 kHz between the $-6$ dB levels (i.e. about 240 kHz when the r.f. input signal takes the tuner into 6 dB of limiting – see Chapter 7 and Fig. 7.19, for example). Moreover, the phase characteristic should be substantially linear over this bandwidth. Lack of phase linearity will result in a multiplex waveform from the f.m. detector similar to that shown in Fig. 9.5(*d*), Fig. 9.6(*b*) or, perhaps, Fig. 9.6(*c*).

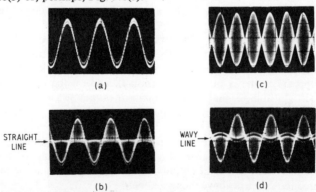

*Fig. 9.5.* Signals monitored in a stereo decoder (Mullard Ltd).

## STEREO RADIO WAVEFORMS

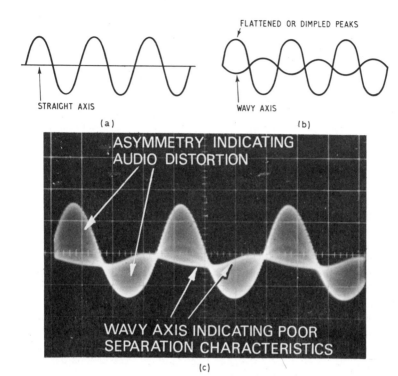

*Fig. 9.6.* Display (*a*) indicates tuner with good frequency and phase response, while display (*b*), with wavy axis, indicates poor frequency and/or phase performance. A tuner producing this waveform would not be very suitable for connecting to a stereo decoder. Waveform (*c*) is direct from the screen of an oscilloscope displaying the stereo multiplex signal, which exhibits bad audio distortion as well as poor stereo separation characteristics.

However, excessive bandwidth with response side skirts not falling away very swiftly will seriously impair the selectivity, and with the increasing exploitation of Band II (the f.m. band from about 88 to 108 MHz) the need for good selectivity is becoming more and more important. Although f.m. channels are 200 kHz wide, there is the likelihood during enhanced propagation conditions for the more distant stations, sometimes separated from the service area regional station by less than 200 kHz, to cause interference, particularly on stereo, when the tuner lacks sharp i.f. channel selectivity.

**Stereo birdies**

One result of this is so-called 'stereo birdies', which is a high-pitched 'twittering' background on the tuned stereo programme. The cause is that an

167

adjacent channel station gets through the i.f. filters so that the f.m. detector receives two signals, the required one and the adjacent channel one. If the two signals are separated by, say, 200 kHz, then the intrinsic non-linearity of the f.m. detector will cause them to beat together and thus yield a beat signal of 200 kHz. This, of course, is well above audibility and under mono conditions would not affect the reproduction.

However, when the wanted signal is stereo-encoded the stereo decoder is activated and harmonics of the 38 kHz switching (i.e. subcarrier) frequency are generated. Since the switching signal is quasi-square-wave the yield is odd-numbered harmonics at 114 kHz (third harmonic), 190 kHz (fifth harmonic), 266 kHz (seventh harmonic), etc. In the case exampled, the 200 kHz detector beat signal will itself beat with the 190 kHz fifth harmonic of the switching signal, thereby producing a 10 kHz signal. This is very high-pitched but within the audio passband, and since it is perturbed by the modulation it manifests as a disturbing 'warbling' on the background of the wanted programme. Clearly, other beat combinations are possible from stereo stations at different spacings.

Under extreme conditions even tuners with good i.f. channel selectivity will pass adjacent channel signal in varying degrees, and contemporary tuners combat this possibility by the use of low-pass filtering between the f.m. detector and stereo decoder. The filter takes a sharp dip at the h.f. end of the stereo multiplex (around 53 kHz), which thus prevents – or significantly discourages – adjacent station beats from reaching the stereo decoder. However, poorly designed filters and/or excessive phase shift can impair the stereo separation.

The nature of the switching signals and any beat tones can be detected with the oscilloscope, and in this respect it is instructive to examine the various signals present in the phase-locked type of decoder. It is noteworthy that the frequency-multiplex type of decoder is less vulnerable to birdies interference than the time-multiplex decoder in any of its guises.

F.M. front-end selectivity is also of increasing importance. Lack of selectivity here reduces the ability of the tuner to handle a multiplicity of strong aerial signals without interference from unwanted stations or spurious responses. The author has devised a test for front-end selectivity 'figure of merit', and this is explained in *The Audio Handbook*. Clearly, the large signal handling ability of the front-end transistors (r.f. amplifier and mixer) is also important in this respect, and the trend is towards the use of field effect transistors (since these have essentially square-law characteristics and hence are significantly less prone to second-harmonic components than bipolar transistors) and bipolar r.f. amplification with the transistor running at relatively high collector current.

More information on the audio side of f.m. tuners is given in the next chapter.

# 10: TESTING AUDIO EQUIPMENT

A POTENT combination for testing audio equipment is the oscilloscope partnered with an audio signal generator capable of delivering sine and square waves over the frequency range 20 to 20,000 Hz. The generator provides the input test signal and the oscilloscope displays the waveform of this at any point in the system, thereby enabling the waveform at the input to be compared with that at the output. A dual-trace oscilloscope, giving two simultaneous displays, facilitates waveform comparison checks of this kind, since the applied input signal can then be fed to the Y1 input and the output signal to the Y2 input. Controls on the scope allow the resulting two waveform traces to be superimposed as nearly as possible, one upon the other, and any distortion on the output waveform is clearly revealed by this not 'fitting' exactly over the top of the input waveform.

On the other hand, valuable tests are possible with a single-trace scope if the quality of the input signal is a known factor. For instance, many audio generators deliver low-distortion sine-wave signals (often less than 1 per cent and sometimes less than 0·1 per cent), so any visible distortion on the output sine-wave display signifies conclusively that all is not well with the amplifier under test if it is one specified as having a low distortion factor (i.e. a hi-fi amplifier).

However, many medium quality audio channels in radios and television sets, tape recorders and inexpensive sound reproducing equipment produce distortion rising as high as 10 per cent at full output, and this is clearly revealed by the display deviating from the true sine-wave shape.

**Spurious signals**

The oscilloscope without a generator will check the presence of spurious signals in the amplifier channel at any frequency within the Y bandwidth, for it sometimes happens that an amplifier or audio section develops a fault that results in h.f. oscillation outside the audible spectrum which cannot be heard in the loudspeaker. The performance of the amplifier is adversely affected, however, with distortion symptoms which, without the oscilloscope, might well be put down to some other cause, such as faulty biasing, output transformer, low h.t. voltage and so forth.

H.F. signal shows on the horizontal trace – with the amplifier under test

# SERVICING WITH THE OSCILLOSCOPE

quiescent – as bursts of sine-wave signal (see page 175). Usual causes of this trouble are open-circuit bypass and decoupling capacitors, poor screening between the output and input circuits, defective design, inadequate output transformer, phase-shift in the coupling between stages or in the negative feedback loop, and excessive negative feedback.

It is sometimes necessary to signal-drive the amplifier before spurious h.f. oscillation is produced. The drive signal then tends to 'trigger-off' the oscillation, so to speak. This is where the signal generator comes in useful again, although it is possible to drive the amplifier with ordinary programme signal and monitor this at the output. H.F. bursts are clearly revealed as 'bulges' on the audio waveform display.

## Gain

With a signal generator whose set output voltage is known (i.e. an instrument with a calibrated signal output control), the overall gain of the amplifier or the gain of any stage can be assessed by using the scope to monitor the signal output voltage and to see how many times greater this is than the input voltage. Even if the generator is not calibrated in terms of output signal voltage, it is still possible to establish gain by setting the vertical trace amplitude a fixed number of centimetres on the graticule, as obtained from an *input signal*, and then transferring the Y input to the output of the stage being measured for gain (at the final output of the amplifier for overall gain measurement). The trace will now be deflected vertically well off the screen due to the stronger signal. Reducing to the original amplitude by switching in Y attenuation gives the gain of the amplifier or stage since this is equal to the attenuation.

It is important to make both the initial and the final Y attenuator adjustments so that the peak-to-peak amplitude of the trace is related to the original fixed number of centimetres on the graticule. Remember also that the gain in decibels is valid only when both the input and output signals are measured for amplitude across the same (or nearly the same) impedance or resistance load.

The voltage gain in simple ratio is given when, say, the input is applied across 100 k and the output monitored across 15 ohms (the secondary of the output transformer, for instance), but this cannot truly be converted directly into decibel gain. It is a different matter, of course, when the output is taken from a high-impedance load as well as the input being applied across one, or when the input is applied across a low-impedance load (say the primary of a microphone transformer) and the output also taken from across a similar load (say the secondary of the output transformer).

## Decibels

An illustration might help. Consider an audio amplifier delivering 15 watts across a 15-ohm load. The voltage will also be 15 volts. This might be obtained from a 100 mV input across 100 k, giving a voltage gain of 150 times. However, the signal voltage across the *primary* of the output transformer might well be 200 volts (or more), and relative to this high impedance point the voltage gain would be 2,000 times! One has to be careful when expressing gain to make sure that the effect of the load impedance has been taken into account, whether the result is expressed in a simple ratio or in decibels.

When the decibel (dB) is used, it must be ascertained that the load resistances $R$ in which the currents $I$ and voltages $E$ operate are equal. When this is so, the following expressions are valid:

$$N \text{ dB} = 20 \log_{10} \frac{E_1}{E_2} \text{ and}$$

$$N \text{ dB} = 20 \log_{10} \frac{I_1}{I_2}$$

When the loads are unequal, the expressions become:

$$N \text{ dB} = 20 \log_{10} \frac{E_1}{E_2} + 10 \log_{10} \frac{R_2}{R_1} \text{ and}$$

$$N \text{ dB} = 20 \log_{10} \frac{I_1}{I_2} + 10 \log_{10} \frac{R_2}{R_1}$$

where $R_1$ and $R_2$ are the loads from which $E_1$, $E_2$, $I_1$ and $I_2$ are taken.

For *power* ratios, the expression is:

$$N \text{ dB} = 10 \log_{10} \frac{P_1}{P_2}$$

where $P$ is the power in watts.

**Tape-recorders**

The scope is useful for checking the waveform of the h.f. bias and erase signal in tape-recorders, and also for checking the amplitude of the signal at the recording and erase heads. The scope applies very little loading to these circuits and the true signal amplitude is given. It is best to convert this from peak-to-peak to r.m.s. value (by multiplying *half* the p–p value by 0·707). The signal current waveform can be appraised by monitoring the h.f. signal developed across a low-value resistor connected in series with the signal current, preferably at the 'cold' end of the circuit. Many recorders incorporate such a resistor in the head circuits for this purpose and for facilitating the measurement of the head currents on a valve voltmeter.

Since the Y amplifier of the vast majority of scopes is flat over the whole of the audio spectrum, it can be used as an output signal indicator in conjunction with an audio generator for checking the frequency response of an

## SERVICING WITH THE OSCILLOSCOPE

amplifier or tape recorder. The overall frequency response of a recorder can be checked by recording sine-wave signal on a virgin tape at intervals over the entire spectrum, starting, say, at 20 Hz and going up to 20,000 Hz. Each frequency recorded can be identified by an announcement recorded on the tape or a table of the frequencies recorded can be used in conjunction with the tape. It is best to start with a fairly long recording at 1,000 Hz as a reference, and it often pays to record 1,000 Hz between each frequency recording. This allows immediate reference to the 1,000 Hz level. Of course, the signal from the generator should remain at a constant amplitude at all the frequencies recorded, and the recording level should be adjusted initially so that over-recording does not occur at any frequency.

Inexpensive recorders pose problems in this connection because their design often incorporates excessive treble boost in the recording amplifiers to maintain a good treble output on replay. This means that while the recording level (as shown on the recording-level indicator of the recorder) will be correct up to about 5,000 or 6,000 Hz, over-recording may occur above that frequency in spite of the input signal being kept at a constant level.

It may be possible to drop the recording level at 1,000 Hz to compensate, but with some machines the level indicator will have to swing off the scale at the higher frequencies. This may not necessarily mean that the tape is, in fact, being over-recorded, since the treble boost is given to outweigh losses which impair the recording of the higher frequencies anyway.

The following tones are most useful: 20, 30, 40, 60, 80, 100 Hz and then to 1,000 Hz at 100 Hz increments, and from 1,000 Hz to 20,000 Hz at 1,000 Hz increments. See also 'Audio sweeping with oscilloscope' later in the chapter.

### Amplifier drive testing

Chapter 2 details the application of the scope for audio amplifier testing, and there an example is given for checking the power output of a hi-fi amplifier. The scope is also extremely useful for checking the drive signal to the power amplifier.

The drive waveform to the output stage can be monitored on the scope at each (assuming a push-pull output stage) valve or transistor input. If clipping commences at the same level as clipping of the output waveform, the trouble lies either in the phase splitter or preamplifier stage. The waveforms in Fig. 10.1 show at (*a*) asymmetrical clipping at the input of one valve and at (*b*) similar distortion but on the opposite half-cycle at the input of the other valve. The trouble in this case was traced to incorrect biasing of a cathode-coupled phase splitter. This was d.c. coupled to the preamplifier valve, from whence it effectively derived its bias.

For a given output power, valves and transistors require a certain r.m.s.

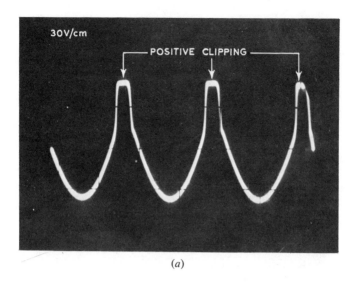

(a)

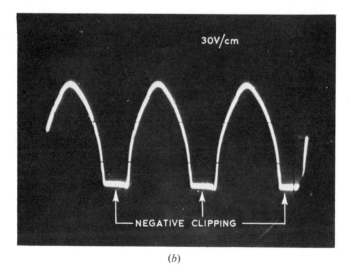

(b)

*Fig. 10.1.* Three examples of clipping on audio drive signals fed to push-pull output valves. Note that the asymmetrical clipping on (*a*) is on the positive half-cycle while that on (*b*) is on the negative half-cycle; poor biasing of the phase-splitter valve was responsible. Shown at (*c*) is an example of symmetrical clipping.

SERVICING WITH THE OSCILLOSCOPE

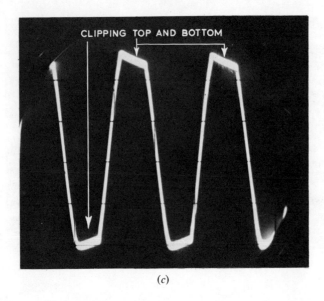

(c)

input voltage. A pair of EL34s, for instance, running in a distributed-load circuit need an r.m.s. drive of the order $2 \times 25$ V for an output of about 20 W. That is, 25 V at each grid with respect to earth or 50 V from grid to grid. This drive has to be distortion-free: if it is and the signal power for a non-clipped waveform across the load is below the rated output, trouble lies in the output stage proper. This may be in the valves, in the biasing or even in the output transformer (i.e. shorting turns), assuming that the h.t. voltage is correct. Simple d.c. tests will show whether the output valves are dissipating as they should.

The waveforms in Fig. 10.1 show a p-p amplitude of about $2\frac{1}{2}$ cm with the input at 30 V/cm. Thus the p-p amplitude is 75 V and the r.m.s. value about 26 V. With asymmetrical clipping of this magnitude, this drive would push the amplifier well into harmonic distortion towards full output!

Symmetrical clipping (Fig. 10.1(c)) implies either that the h.t. is low or that the stage is being symmetrically overloaded. Asymmetrical clipping, on the other hand, signifies either incorrect bias, causing the valve to cut-off on negative peaks, or saturation, resulting from a low-emission valve or limiting effects on positive half-cycles. Grid current on positive peaks and low h.t can also influence the effect.

While the vast majority of contemporary hi-fi amplifiers utilise transistors (and sometimes integrated circuits as well), there are a large number of valve amplifiers still in use, and some enthusiasts would have nothing else! Moreover, at the time of preparing this new edition at least one British

# TESTING AUDIO EQUIPMENT

manufacturer (Lowther Acoustics Ltd) is still making valve amplifiers, which go well with this firm's horn loaded loudspeakers of high efficiency, and even while this text is being written there has been an announcement of a Japanese manufacturer going back to valves (Lux, distributed by Howland-West Ltd).

However, this must not be taken to imply that transistors for hi-fi applications have failed. This is certainly not the case. Indeed, there is a possibility that field effect power transistors will eventually take over from power bipolars, the former having square-law characteristics that are closer to those of valves than bipolars. This means that third and odd-order harmonic distortion is reduced while second and even-order harmonic distortion is essentially cancelled by carefully balanced push-pull operation. Thus the negative feedback required for a given distortion performance can be less with f.e.t.s than bipolars, and this tends to enhance stability and minimise transient intermodulation distortion. Already amplifiers using f.e.t. power transistors have been designed.

**Parasitics**

The oscilloscope can show up any parasitic oscillation in an amplifier, as we have seen. This may be of a frequency removed from the audio spectrum yet still cause havoc to the quality of reproduction. Fig. 10.2 shows how this unwanted signal looks on a sine wave. This was taken across the output load at 100 Hz and with the Y input set to 10 V/cm. Fig. 10.3 shows expansion applied to give greater detail to the spurious oscillation, which actually takes the form of sine waves on the wanted signal waveform.

Parasitic oscillation often results from the use of a reactive load across the

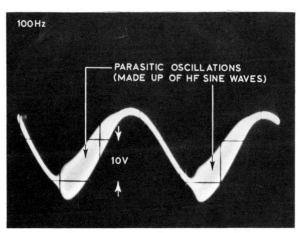

*Fig. 10.2.* Parasitic oscillation on a sine wave output signal.

175

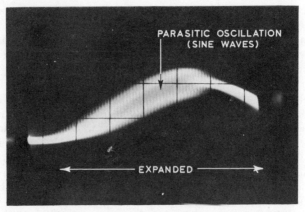

*Fig. 10.3.* Expansion revealing the make-up of the parasitics.

output. Some amplifiers, for instance, do not like electrostatic speakers which offer a capacitive load. Other causes are excessive phase shift in the output transformer, coupled with too great a value of negative feedback, or even failure of decoupling elements, or phase correcting artifices, in the circuit proper. The oscillation shown is, of course, h.f., but low-frequency effects can arise causing the trace at the output to fall and rise vertically on the screen at the frequency of the parasitic.

Almost all amplifiers tend to develop parasitics if too much negative feedback is employed. This happens towards (or usually outside) the audio fringes due to the feedback going positive from phase shift in couplings or transformers. Indeed, one test for amplifier stability is to find by how many decibels the feedback can be *increased* before parasitics appear.

**Phase measurement**

The oscilloscope can be used to measure the phase angle between two equal-frequency sine-wave signals to an accuracy of about $\pm 1$ deg. when the X and Y channels have identical phase characteristics and when a specially prepared graticule is used in front of the screen. This needs to be accurately calibrated and care has to be taken over the possibility of parallax error. However, it is possible to obtain realistic phase comparisons using the ordinary square-ruled graticule.

The reference signal is normally fed to the X input and the phase-shifted signal to the Y input, the relationships then being

$X = \sin \omega t$
$Y = \sin (\omega t + \varphi)$

# TESTING AUDIO EQUIPMENT

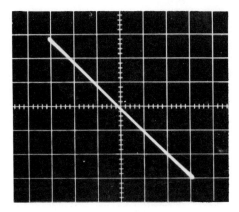

*Fig. 10.4.* A straight line inclined by 45 deg. signifies zero phase difference between two sine-wave signals applied separately to the X and Y inputs, with the timebase switched off, with the signal levels balanced and with identical X and Y channel phase characteristics (see text).

When the controls are adjusted to give equal vertical and horizontal deflections a straight diagonal (i.e. 45 deg.) line is displayed when $\varphi$ = zero phase difference, as shown in Fig. 10.4. This opens into an ellipse and eventually appears as a circuit at 90 deg. phase difference, as shown in Fig. 10.5.

Fig. 10.6 shows a phase difference between 0 deg. and 90 deg., and from this ellipse can be calculated the phase difference, such that sin $\varphi$ = $B/A$. Here the ratio is about 0·6, whose sine is about 37 deg. Hence the phase difference between the two signals represented by Fig. 10.6 is about 37 deg.

The 'reciprocal' angles also apply, of course, so that a 45 deg. line inclined the opposite way to that of Fig. 10.4 could represent the antiphase condition (i.e. 180 deg.), while an ellipse tilted the opposite way to that of Fig. 10.6 could represent a phase difference of 143 deg. (i.e. 180 deg. −37 deg.).

As already mentioned, the phase characteristics of the X and Y amplifiers must be identical, and the levels of the X and Y signals (with the scope's timebase switched off of course!) must be equal so that in the zero phase-difference condition the straight line is exactly at 45 deg. inclination.

The nomograph below allows the phase angle difference to be obtained directly from different *A* and *B* parameters. A straight edge is arranged to

177

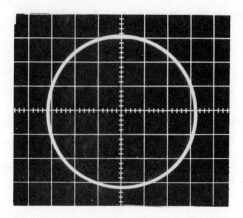

*Fig. 10.5.* At 90 deg. phase difference the 45 deg. straight line opens to form a circle as this oscillogram shows.

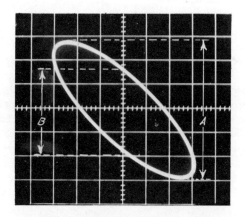

*Fig. 10.6.* By measuring parameters *A* and *B* and referring them to the nomograph on page 179 the phase difference can be directly obtained. The sine of the phase difference is equal to *B/A*, and in this example the phase angle works out to about 37 deg. since ratio *B/A* is about 0·6.

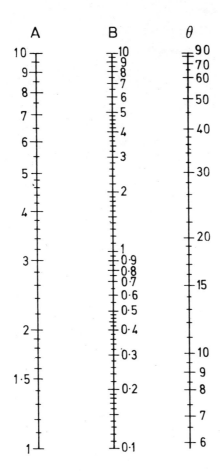

Nomograph for direct determination of the phase differences from the parameters $A$ and $B$ in Fig. 10.6. (After a nomograph devised by J. F. Golding of Marconi Instruments Ltd.)

cut the displayed $A$ and $B$ parameters and the angle indicated by the straight edge corresponds directly to the phase difference.

Fig. 10.7 shows the effect of symmetrical clipping on a phase pattern due to amplifier overdrive.

SERVICING WITH THE OSCILLOSCOPE

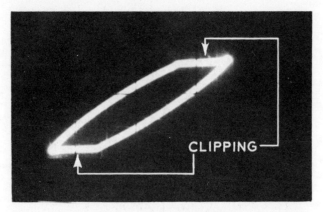

*Fig. 10.7.* This is the effect of symmetrical clipping, as illustrated by Fig. 10.1(*c*), on a phase trace.

**Frequency response**

The best way of checking the frequency response of an amplifier is to apply sine-wave signal from an audio oscillator, whose frequency is adjustable over about 20 Hz–20 kHz, to the amplifier input at relatively low impedance (that is, the oscillator source impedance *circa* 600 ohms), turn the volume control to maximum, switch off filters, set the tone controls for 'flat' and adjust the oscillator signal level for an output of about 1 W (depending on the power of the amplifier). With the input signal being kept at a constant level at all frequencies, the differences in level at the output, relative to 1 kHz, can be plotted in decibels against frequency on a graph.

The effect of the l.f. and h.f. filters, tone controls, equalised stages, loudness filter, etc. can be seen by replotting with these items active or in circuit. This time-consuming exercise can be lightened by using a sweep-frequency oscillator and viewing the results on the screen of an oscilloscope, as explained later.

**Power response**

The power response can be plotted similarly, but this time the plot is made at the rated power of the amplifier (all channels driven with multi-channel amplifiers) with reference to a given level of harmonic distortion (1 per cent DIN, but 0·5 per cent is better). Thus the plot is that of the frequency spectrum over which the amplifier can yield its rated power without the distortion rising above the reference level.

TESTING AUDIO EQUIPMENT

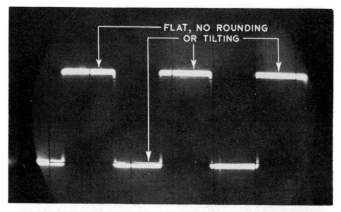

*Fig. 10.8.* 1,000 Hz square wave at the output of 20 W amplifier with the input applied to the power amplifier, not the preamplifier.

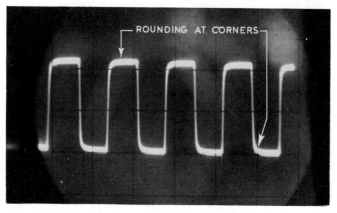

*Fig. 10.9.* When the frequency of the square wave is increased, the fall in h.f. response is evidenced by rounding corners, as this display shows.

**Square-wave testing**

A good idea of how an amplifier is behaving frequency-wise can be gleaned by the use of square waves. This is because a square wave is composed of the fundamental frequency and a multiplicity of odd harmonics of the fundamental added in specific phase relationship and amplitude (Chapter 1). The more harmonics there are, the better the square wave. The fundamental frequency and the harmonic components, of course, are pure sine waves. Thus, if an amplifier handles a square wave without distortion it follows that it

181

SERVICING WITH THE OSCILLOSCOPE

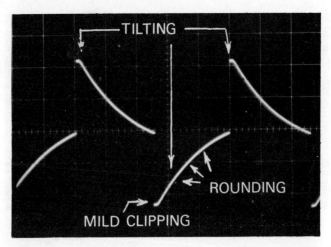

*Fig. 10.10.* The tilt indicates l.f. attenuation and the curvature or rounding phase shift. The same direction of tilt can also signify h.f. boost and phase-lead shift.

must be passing all the harmonic components and therefore has a good bandwidth. Square-wave testing tells other things, as we shall see.

Fig. 10.8 shows a 1,000 Hz square wave at the output of a 20 W power amplifier with the signal applied to the power amplifier input (*not* the preamplifier input). Points to note are the freedom from corner rounding and ringing and lack of tilt at the top and bottom of the waveforms. At 20,000 Hz the waveform looks like Fig. 10.9. Notice the tendency towards rounding of the leading and trailing corners. This implies a falling h.f. response *relative to the repetition frequency*.

To maintain perfect square-wave definition, an amplifier has to have a response up to about ten times the repetition frequency – 10,000 Hz for Fig. 10.8 and up to 200,000 Hz for Fig. 10.9. Fig. 10.9, in fact, shows that the response is not up to that standard but that is not surprising in an audio amplifier. Actually, the waveform at 20,000 Hz is remarkably good. Some amplifiers start rounding like this at 2,000 Hz – yet still *sound* good.

Poor response at the low-frequency end of the spectrum is indicated by tilting of the top and bottom of the waveform as in Fig. 10.10. This was taken at 80 Hz and the tilt is considerably aggravated by switching in the high-pass filter when the input is applied to the preamplifier.

Fig. 10.11 shows 'ringing' instigated by overshoot at the start of the flat trace (i.e. directly following the swiftly changing transient). A display like this when the amplifier is loaded resistively indicates a tendency for instability. A degree of overshoot is not uncommon when a transistor amplifier is

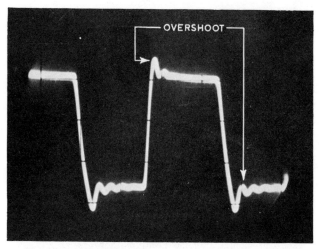

*Fig. 10.11.* Overshoot due to action of negative feedback.

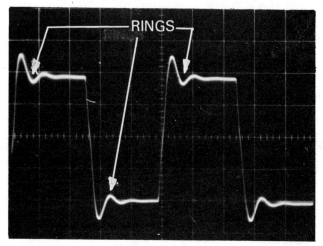

*Fig. 10.12.* Rings on a 10 kHz square wave at the output of a transistor amplifier loaded with 8 ohms in parallel with 2 μF.

loaded with capacitance in shunt with resistance, and a typical 10 kHz display of this kind is shown in Fig. 10.12, where the load consisted of 8 ohms in parallel with 2 μF.

Excessive h.f. attenuation in an amplifier eliminates most of the harmonic

SERVICING WITH THE OSCILLOSCOPE

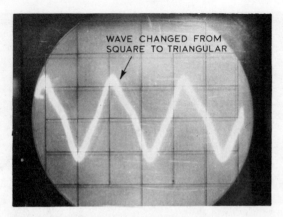

*Fig. 10.13.* Excessive attenuation of the harmonic components of a square wave leaves mainly the fundamental component only, as shown here.

components of a square wave and severely distorts the display, as in Fig. 10.13.

The time taken for a square wave (or transient pulse) to rise from 10 to 90 per cent of its final value is the rise time. Fig. 10.14 illustrates this on a sweep of 2 μs/div. About 3·2 divisions are covered, which puts the rise time at $2 \times 3·2$ or about 6·4 μs. Also see Chapter 1 under 'Rise time' and Fig. 1.4(*a*).

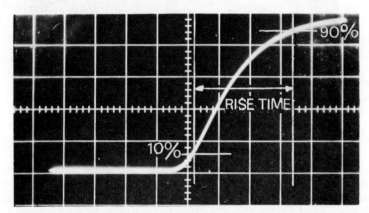

*Fig. 10.14.* Illustration of rise time on a sweep of 2 μs/div. Here the rise time is approximately 6·4 μs. See text and also Chapter 1 under 'Rise Time' and Fig. 1.4(*a*).

## TESTING AUDIO EQUIPMENT

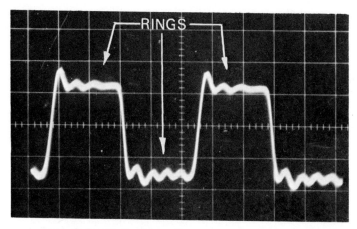

*Fig. 10.15.* Slight ringing effects on square wave output.

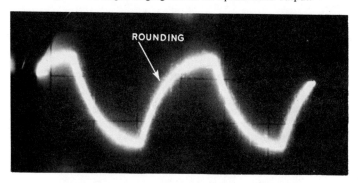

*Fig. 10.16.* When a square wave is applied to a pick-up input (equalised), the output display is as shown here. The rounding is due mainly to bass boosting, the converse of which is treble cut.

Square-wave testing reveals any tendency to *ringing* (damped oscillation) in the amplifier. Fig. 10.15, for example, shows just a little ringing on the horizontal parts of the trace. This is due to oscillation in the loss inductance of the output transformer. Small ringing like this has no effect on performance but excessive rings can give overhang on the reproduction. Amplifiers adopting inductive equalisation or filter components are more prone to this trouble if the circuits are not sufficiently damped.

When a square-wave signal is applied to the input of the preamplifier a great deal of difference at the output should not be observed provided the signal is applied to a 'flat' input. However, Fig. 10.16 shows the output result when the signal is applied to the equalised pick-up input. The equalisation applies bass boost and, to the square wave, this appears like treble cut. Hence the rounded features of the waveform.

### Effect of tone controls on signal

The series of oscillograms in Fig. 10.17 shows the effect of tone controls on a square-wave signal; (*a*) shows the output with the bass and treble controls carefully adjusted to the 'flat' condition (not always as indicated on the scales), (*b*) and (*c*) show bass boost and cut respectively (with treble control 'flat'), while (*d*) and (*e*) show treble boost and cut respectively (with bass control 'flat'). The square wave has a repetition frequency of 1 kHz.

The controls are carefully adjusted in conjunction with each other, ignoring scale markings, until the nearest approach to (*a*) is achieved. One can then be assured that the controls are not far from their 'flat' settings.

### Hum test

One way of checking for hum is to switch the X base off and apply to the X-input terminal a 50 Hz signal (possibly from the oscilloscope's calibration socket). The X input should then be adjusted to give a little over

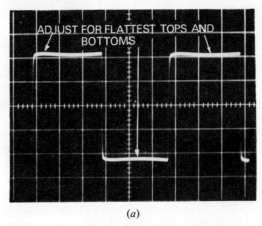

(*a*)

*Fig. 10.17.* This series of waveforms (*a*) to (*e*) shows how the tone controls of a hi-fi amplifier can be adjusted to the 'flat' position. At (*a*) is shown the output waveform when the controls are very carefully adjusted for zero tone control effect (i.e. the 'flat' position); (*b*) and (*c*) bass boost and cut respectively; and at (*d*) and (*e*) treble boost and cut respectively with bass control 'flat'. Input frequency 1,000 Hz.

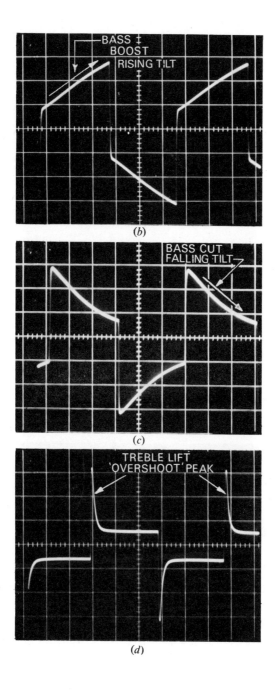

## SERVICING WITH THE OSCILLOSCOPE

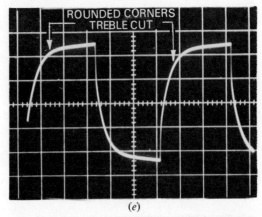

(e)

half-screen deflection. The Y input should be connected to the output of the amplifier (across a correct value load) and the switched input should be loaded with a matching value resistor.

Now, when the amplifier is switched on and the volume control advanced, a display, like Fig. 10.18, showing hum at the mains frequency, or Fig. 10.19, showing hum at twice mains frequency, may be seen. The former would, of course, be obtained by direct pick-up of mains field, such as from the heater wiring or mains-carrying transformer or cable, while the latter would be h.t. supply ripple from a full-wave rectifier circuit. Excessive hum from this cause could indicate insufficient reservoir or smoothing capacitance or shorting turns in the smoothing choke.

The 'grass' effect on Fig. 10.18 is noise in the early stages of the amplifier, these running at maximum gain minus input signal. The hum component here

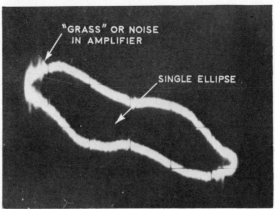

*Fig. 10.18.* Check for hum at mains frequency (see text).

# TESTING AUDIO EQUIPMENT

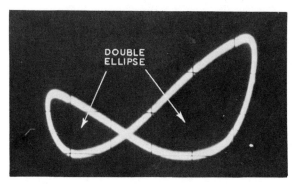

*Fig. 10.19.* Check for hum at twice mains frequency (i.e. hum on h.t. line with full-wave h.t. rectifier) – see text.

is very small indeed, being no greater than inherent noise level. Fig. 10.19, though, demanded electrolytic capacitor replacement to effect a cure.

**Testing for low-level distortion**

Although high-level distortion is visible on the displayed waveform, as we have seen, low-level distortion gives so little deformation of the wave display

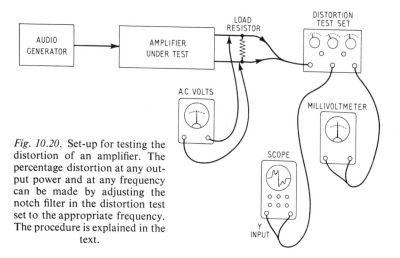

*Fig. 10.20.* Set-up for testing the distortion of an amplifier. The percentage distortion at any output power and at any frequency can be made by adjusting the notch filter in the distortion test set to the appropriate frequency. The procedure is explained in the text.

189

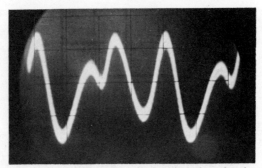

*Fig. 10.21.* The nature of the distortion signal generated by a class B audio amplifier. This corresponds to about 0·2 per cent at 1,000 Hz.

that it cannot be observed by the direct oscilloscope method. It is necessary, therefore, to adopt a more sophisticated testing method, and Fig. 10.20 shows one that is often used. Here the test amplifier is fed with a sine-wave signal from an audio generator having a very low distortion characteristic (about 0·01 per cent or less) and the output is loaded with a non-reactive resistor – instead of the loudspeaker – capable of handling the full average power. The audio voltage $V$ across this load $R$ is read on an a.c. voltmeter and the power output $W$ calculated by using the expression $W = V^2/R$, where $W$ is in watts and $R$ is in ohms.

The signal across the load is also fed into a distortion test set, which is basically a device having a tunable 'notch filter' giving a very high attenuation (– 80 dB or more) to the fundamental frequency. The signal delivered by this test set is thus the distortion components only, the fundamental signal having been deleted by the notch filter. The distortion components represent the total harmonic distortion plus noise (i.e. distortion factor), and these are measured on the millivoltmeter. By comparing the distortion signal amplitude with the amplitude of the signal across the load, the distortion factor percentage is obtained. For instance, if the millivoltmeter registers an output 60 dB down from that on the a.c. voltmeter, then the distortion is 0·1 per cent (1,000 times down). Similarly, a difference of 40 dB would indicate a distortion of 1 per cent (100 times down).

The scope can be connected to the test set, as shown, to reveal the nature of the distortion signal, indicating whether it is second, third, odd or even-harmonic, and an oscilloscope display of this kind is given in Fig. 10.21. This represents odd-numbered harmonic distortion at a level of about 0·2 per cent at 1,000 Hz, and is the kind of distortion generated by class B hi-fi transistorised amplifiers. Class A amplifiers tend to generate less odd-numbered harmonic distortion or, at least, fewer higher-order odd-numbered harmonics than class B amplifiers.

## TESTING AUDIO EQUIPMENT

A dual-trace oscilloscope is useful when checking distortion factor with the set-up shown in Fig. 10.20, since the output signal directly across the load can be displayed on one trace and the distortion on the other. Since the distortion contains not only all the harmonic components after the fundamental has been removed, including noise components, the residual is the distortion factor, and the distortion test set in Fig. 10.20 is more correctly a *distortion factor* meter or unit.

The oscillograms in Fig. 10.22 were obtained with a dual trace oscilloscope. Oscillogram (*a*) indicates that the distortion is essentially second harmonic since there are twice as many distortion waves as there are direct signal waves. This sort of distortion is more palatable than odd-order harmonic distortion. In fact, the display here is representative of a first class hi-fi amplifier, and since the voltage of the distortion is 60 dB below that of the direct signal, the distortion is a mere 0·1 per cent.

The distortion at (*b*) is even lower, being about 0·03 per cent. The distortion here is virtually down to the noise level of the amplifier, the noise components being indicated by the ragged nature of the residual.

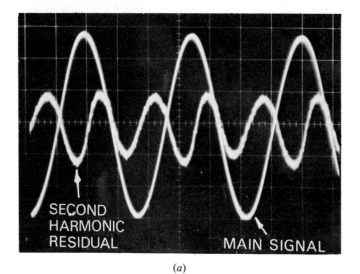

(*a*)

*Fig. 10.22.* Distortion waveforms at 1 kHz and 10+10 W into 8-ohm loads on dual trace oscilloscope, with one trace showing the main signal and the other trace showing the distortion residual with high Y gain: (*a*) essentially second harmonic distortion; (*b*) distortion virtually down to amplifier noise level; (*c*) traces of odd-order harmonics plus noise; (*d*) crossover distortion. See text for more details.

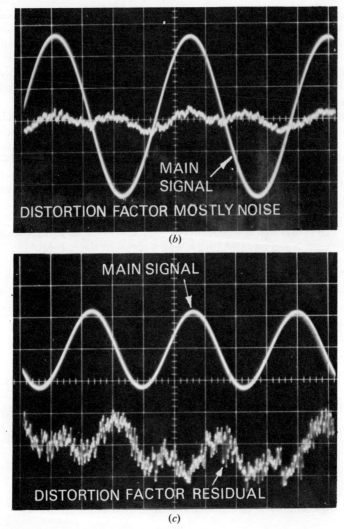

(b)

(c)

Oscillogram (c) also shows very low level distortion (*circa* 0·015 per cent), but this time there are a few odd-order components and, of course, slightly higher relative noise level.

The display at (d) shows crossover distortion due to switching discontinuity of the push-pull output transistor pair. This distortion generates high-order odd harmonics which, at high level, are singularly unpalatable to the critical ear.

# TESTING AUDIO EQUIPMENT

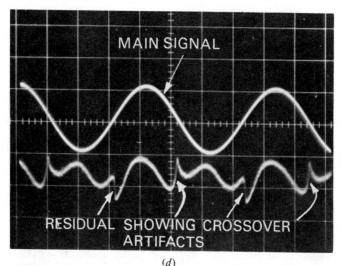

(d)

## Slewing rate

The rise time of an amplifier gives its response to both h.f. sinusoidal and transient pulses, *provided that the maximum slewing rate of the amplifier is not exceeded*. The maximum slewing rate is expressed in volts per second (V/s or, more usually, V/µs). It is the highest frequency at which the amplifier will deliver its rated voltage before being limited by the slewing rate, the limit resulting from some stage in the amplifier being incapable of yielding sufficient current to charge the capacitance of its load, such that

$$i = C \frac{de}{dt}$$

When an amplifier is driven by a sine wave of $e_i = E \sin \omega t$, maximum $de/dt$ obtains at values of $\omega t$ corresponding to zero, $\pi$, $2\pi$, etc. radians. Thus

$$\frac{de_i}{dt} = \omega E \cos \omega t$$

where $E$ is the maximum voltage and $\omega$ is $2\pi f$.

Evaluation gives $de/dt$ (or slewing rate SR) as

$$SR = E\, 2\pi f$$

where $f$ is the highest frequency (in hertz) at which the rated *peak* output voltage $E$ can be obtained.

Thus one way of measuring the slewing rate is to drive an amplifier with sine-wave signal to its rated peak output voltage (hence rated output power into its rated load) while increasing the frequency of the input sine wave

## SERVICING WITH THE OSCILLOSCOPE

until the distortion begins to increase swiftly. That frequency and the *peak* voltage across the load can then be used in the above expression to obtain the slewing rate, taking care to avoid amplifier damage due to sustained h.f. power.

It is best to use an oscilloscope to detect the onset of h.f. distortion, and if the scope is calibrated the peak value (not peak-to-peak) of the voltage across the output load can at the same time be obtained. For example, if the peak-to-peak is 20 V (corresponding to a peak of 10 V and an r.m.s. value of 7·07 V, which would yield 6·2 W into an 8-ohm load) and the highest frequency at which this can be obtained is 50 kHz, then the slewing rate would be 3·14 V/μs.

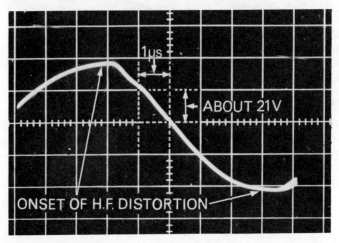

*Fig. 10.23.* Oscilloscope method of determining amplifier slewing rate. See text for a full description.

Fig. 10.23 shows an oscilloscope slewing rate measurement. Here the amplitude is 20 V/div, the sweep 1 μs/div, the frequency 86 kHz and the power into the load (which is 8 ohms) 100 W average sine wave. Thus the r.m.s. voltage across the load is 28·28 V, making the peak voltage virtually 40 V, which is as indicated on the oscillogram. The slewing rate, therefore, works out to about 21·6 V/μs. The oscillogram shows a rate of change of a little over 20 V/μs, which is not very far from the calculated method!

### Audio sweeping with oscilloscope

It is possible to obtain a display of the frequency-amplitude characteristics of an amplifier, network, etc. on the screen of an oscilloscope by the use of a sweeping functions generator. The technique involved is similar to

## TESTING AUDIO EQUIPMENT

that described in Chapter 7 for r.f. response displays and visual alignment. The functions generator produces sine, square and possibly triangular waves from a very low frequency (commonly below 1 Hz) to a frequency well above audio (to 200 kHz or even higher). The frequency can be selected by a calibrated control in the normal way, but in addition the frequency can be caused to change automatically over a ratio typically of 1000:1, the frequency span being selected by a switch.

The functions generator used by the author can be swept from 2 Hz to 2 kHz on the first range, from 20 Hz to 20 kHz on the second range and from 200 Hz to 200 kHz on the third range. The sweep can be achieved either by an internal ramp generator or by an external ramp signal such as delivered by the X output terminal of an oscilloscope. The rate of the sweep when internally obtained is also adjustable over 2·5 ms, 250 ms and 25 s. Facilities are also available to decrease or increase these sweep rate values.

Moreover, the internal sweep can be either linear or logarithmic, the latter being required to avoid compression of the X frequency axis of the display, the display frequency-wise then corresponding to that provided by the usual three-decade graph paper commonly used for plotting frequency-amplitude response characteristics of audio equipment. The generator also delivers a synchronising pulse to trigger the scope at the commencement of a sweep (the alternative, of course, would be to get the timebase signal of the scope to activate the sweep, but the author has found the sync method more convenient, especially when the sweep is required to be logarithmic).

Such a functions generator and a good oscilloscope with positive triggering facilities and single shot button constitute a potent pair of instruments for audio equipment testing and evaluation, the set-up being as illustrated in Fig. 10.24. Fig. 10.25 shows one partnership of such instruments employed in the author's audio laboratory. The functions generator is the top of the two instruments on the left-hand side of the oscilloscope. The bottom instrument is an ordinary low distortion audio oscillator which is sometimes used in conjunction with the functions generator.

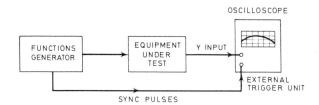

*Fig. 10.24.* Instrument set-up when using a functions generator for audio frequency response displays. Here the start of the scope's sweep is triggered by a starting sync pulse from the functions generator.

195

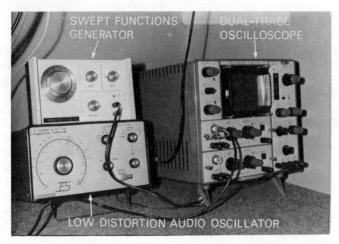

*Fig. 10.25.* The functions generator and oscilloscope used for audio sweep frequency displays by the author.

Clearly, from Fig. 10.24 it will be understood that when the functions generator starts a sweep the sync pulse which it produces triggers the scope so that at the same moment this also starts an X sweep. When the sweep time is adjusted to match that of the functions generator a sine-wave display of increasing frequency is obtained, and it is the envelope of this which gives the frequency-amplitude characteristics of the equipment under test.

To obtain a meaningful amplitude/frequency readout a graticule calibrated in decibels down the Y axis and in frequency along the X axis needs to be used in front of the screen. Thus the scope should be capable of easy graticule change, and if photography is contemplated the graticule must also be illuminated. It is a simple matter to change the graticule of the scope used by the author, and the intensity of illumination is externally adjustable.

The series of oscillograms in Fig. 10.26 gives examples of frequency sweeps so obtained. Each is described in the caption.

It is possible to rectify the signal and then apply the resulting direct voltage to the Y input of the scope, with the scope switched to d.c., and an example of one sweep by this method is given in Fig. 10.27. However, owing to the long time-constant of the smoothing circuit, which has to be active at very low frequencies, the sweep rate needs to be substantially increased (i.e. longer sweep time). The sweep rate for the oscillograms in Fig. 10.26 was 2·5 ms, while for that in Fig. 10.27 it was 25 s, and even then the smoothing was insufficient to eliminate all traces of l.f. signal.

An alternative scheme devised by Fred Judd, Technical Editor of *Practical Hi-Fi and Audio* magazine, has extreme merit since he eliminates

# TESTING AUDIO EQUIPMENT

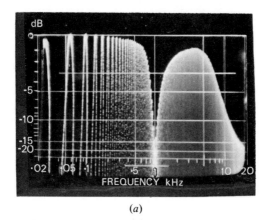

(a)

*Fig. 10.26.* Examples of sweep frequency displays. (a) Notch filter tuned to 1 kHz. (b) F.M. frequency response characteristic. Here the f.m. generator was modulated by the sweep signal via 50 μs pre-emphasis filter corresponding to the de-emphasis of the tuner under test. Notice the −3 dB fall at 10 kHz and the swift decline into the 19 kHz pilot tone notch. (c) Replay response of cassette tape recorder, showing the 19 kHz notch which is now commonly used when Dolby noise reduction is incorporated to prevent the Dolby circuits from being affected by 19 kHz pilot tone signal when recordings are made from stereo radio. (d) Loudness characteristic of hi-fi amplifier. This gives bass boost at the lower volume control settings. (e) Bass and treble tone control responses of hi-fi amplier.

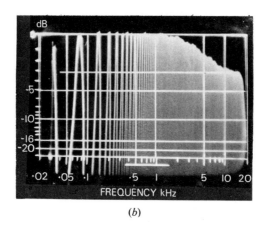

(b)

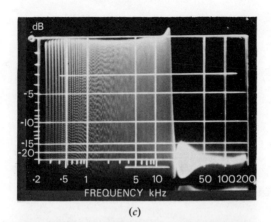

(c)

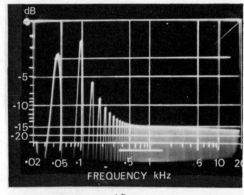

(d)

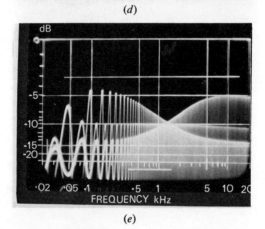

(e)

TESTING AUDIO EQUIPMENT

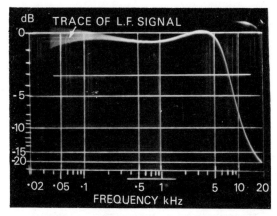

*Fig. 10.27.* Response display when the signal is rectified and the resulting direct voltage fed to the Y input of the scope, with the Y input switched to d.c.

the sine waves of the sweep, leaving only the 'envelope' of the display, by applying 'brightening' pulses to the Z input corresponding to the peaks of the waves. The brightness of the display is then reduced so that only the brightened peaks are seen, as shown in Fig. 10.28. This technique allows

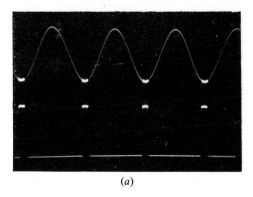

(a)

*Fig. 10.28.* Sweep frequency displays by Fred Judd, Technical editor of *Practical Hi-Fi and Audio* magazine, showing his method of eliminating the sinewaves for a single line display. (a) The tips of the sine waves (upper trace) are brightened by pulses as shown by the lower trace. (b) The brightness is then backed off so that only the tips of the sinewaves are displayed. (c) The resulting effect on a 10 Hz–100 kHz sweep.

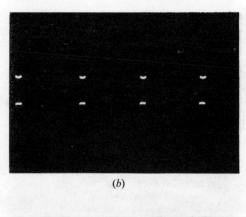

(b)

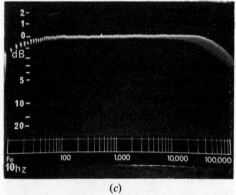

(c)

relatively fast sweeping and hence greater mobility of the beam while at the same time producing a display which is more like the single line response in Fig. 10.27. The mild smudging effect at the h.f. end of the response curve is due to the phase of the brightening pulses deviating from that of the sinewaves.

Another interesting sweep display by Fred Judd is given in Fig. 10.29, which is that of a loudspeaker in a room. The loudspeaker is caused to radiate the swept frequency, and this is picked up by a microphone whose signal is then applied to the Y input of the oscilloscope.

The method can also be used to check the overall recording/replay characteristics of a tape recorder by recording the sweep on one stereo track and the sync pulses on the other, and an example of the sweep of a cassette machine as obtained by the author is given in Fig. 10.30. The 'ragged' nature of the envelope results from noise and mild drop outs.

# TESTING AUDIO EQUIPMENT

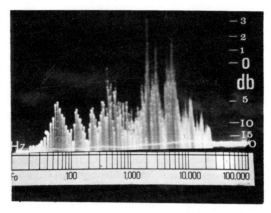

*Fig. 10.29.* Another sweep by Fred Judd, this time of a loudspeaker in a room.

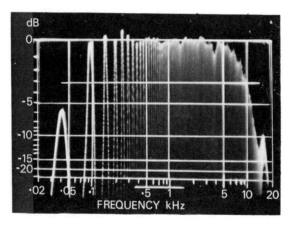

*Fig. 10.30.* Display of overall recording/replay response of cassette tape recorder, with the sweep applied to one stereo track and the sync pulses to the other. Note with all displays of this kind, where the graticule is separate, and slightly displaced, from the cathode-ray tube, account has to be taken of parallax error, and the graticules and camera system used by the author take parallax error into account.

It will have been noticed that in all examples of the sweeps given the dB Y (amplitude) axis is non-linear. That is, there is gradual compression in terms of decibels with reducing amplitude. This sort of response is desirable for some displays because it emphasises small deviations of response about the 0 dB datum. However, for greater dynamic range a linear dB scale is required, and this is obtained either by using a logarithmic Y amplifier for the oscilloscope or by interposing a logarithmic amplifier between the equipment under test and the Y input of the scope, the deflection in both cases then being in terms of d.c. (as in Fig. 10.27).

A final point concerning the responses such as in Fig. 10.26 is that the incomplete horizontal line towards the top of the graticule corresponds to the $-3$ dB point, while the short horizontal line at the bottom is merely a datum line for adjusting the trace to the graticule.

The author is currently employing a precision a.c./d.c. converter and a logarithmic amplifier to provide oscilloscope frequency-amplitude plots that correlate closely with those obtained by hand-plotting on graph paper and by automatic pen recording. The converter/log amplifier is interposed between the swept audio signal from the item under measurement and the d.c. Y-input of the oscilloscope, as shown in Fig. 10.31.

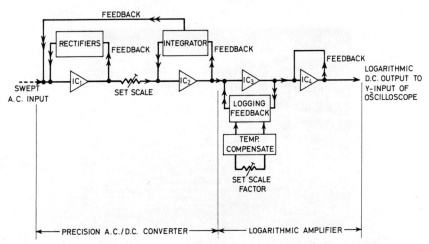

Fig. 10.31. Schematic diagram of precision a.c./d.c. converter and logarithmic d.c. amplifier.

Operational amplifiers are used in both the a.c./d.c. converter and the log amplifier, the devices being chosen for the former to provide a constant and low ripple d.c. from input signal over the range 2 Hz to 200 kHz. The resulting d.c. is then applied to the input of the log amplifier, and the output is used to drive the d.c. Y-input of the oscilloscope. Full-wave rectification in a feedback circuit, followed by a stage of integration, gives the required d.c. with a very low ripple content.

## TESTING AUDIO EQUIPMENT

The d.c. log amplifier utilises a bipolar transistor feedback circuit to provide the logarithmic function. This happens because the relationship between collector current and emitter–base voltage of a bipolar transistor is precisely logarithmic over a usefully wide range. By careful design, suitable choice of operational amplifiers and care with offset adjustments it is possible to secure a dynamic range as wide as 100 dB.

Temperature compensation is required because the resulting log term is directly proportional to the absolute temperature. Without compensation the scale factor would also change with temperature.

For suitable display, the author has developed a graticule compensated for parallax error which, when scaled for 5 dB/div, correlates with the 'standard' graph plots such that the distance over a frequency decade corresponds to 25 dB.

Two example displays are given in Fig. 10.32. At (*a*) the top trace shows the frequency response of an f.m. tuner referred to 50 μs pre-emphasis with only one channel modulated, while the bottom trace shows the level of the signal in the non-speaking channel and hence the stereo separation over the

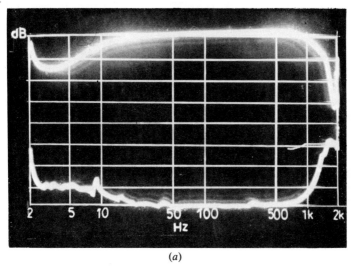

(*a*)

*Fig. 10.32.* (*a*) Frequency plot using an oscilloscope and a.c./d.c. converter with logarithmic amplifier, showing f.m. stereo frequency response of the speaking channel top trace and breakthrough signal in the non-speaking channel bottom trace. Scale Hz × 10 and 5 dB/div. (*b*) Similar display, but this time of a gramophone pickup, where the top trace is the frequency/amplitude response of the speaking channel and the bottom trace the breakthrough signal in the non-speaking channel. The rising low-frequency response is due to constant-amplitude modulation to about 500 Hz and constant-velocity modulation to 20 kHz. Scale Hz × 10 and 5 dB/div.

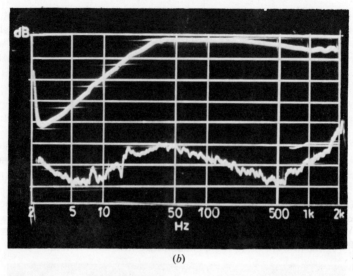

(b)

audio spectrum 20 Hz–20 kHz. This sort of display, of course, also requires the use of a stereo f.m. signal generator whose modulation in either channel can be swept over the band of interest.

The oscillogram at (b) is also interesting since it refers to the frequency response of a gramophone pickup. The top trace corresponds to the output of the speaking channel and the bottom trace to the breakthrough signal in the non-speaking channel, again revealing clearly the stereo separation over the frequency range. The sweep signal in this case was obtained from a gliding-tone test record, the oscilloscope being triggered by a preceding 1 kHz tone. It should be noted that the test record had constant-amplitude modulation to 500 Hz (hence the rise of 6 dB/octave at the low-frequency end) and constant velocity modulation from 500 Hz to 20 kHz.

# INDEX

A.C./D.C. converter, precision, 202
Active probe, 17
Alignment
  f.m. tuner/receiver, 129
  stereo radio, importance of, 166
  sweep method, 115–131
  techniques of, 118
  TV sound channel, 122
Amplifier, Y
  differential, 12
  plug-in, 12
Applications of oscilloscope, simple, 1, 23–32
Audio applications of oscilloscope, 30
Audio equipment, testing in, 169–204
Audio gain, 170
Audio, sweep response displays, 194–204

Bandwidth for f.m. stereo radio, 166
Bandwidth, Y channel, 6
Blanking waveforms, 65
Brightness control adjustment, 15

Capacitor check, TV sound output, 103
Ceefax information (in waveform), 140
Charging time of capacitor, 59
Choosing an oscilloscope, 13
Clipping of signal, 11, 31, 126, 173, 174
Colour bursts, 135
Colour-difference signals, 132
Colour tube, gun assembly, 146
Colour TV
  basic principles, 134
  PAL system, 136
  reception, 137
  signals, 28
  waveforms, 132–157
Compensated voltage-divider probe, 16, 19

Convergence
  dynamic, 145–146
  in colour TV, 143
  static, 145–146
Corona interference, 38, 43

Decibels, 170
Decoder waveforms, stereo, 165
Deflection sensitivity, 2
Delayed sweep, 12
Detector probe for sweep alignment, 121
Differentiator, 58
Digital frequency counter, Heathkit, 131
Displays (oscillograms and TV photographs)
  amplitude modulation, 14
  audio signal, 103
  audio waveforms, 173–203
  blanking pulses, 65–66
  cassette recorder sweeps, 201
  clipping, 11, 31, 126, 173, 174
  colour TV video signals, 150–157
  corona, e.h.t. discharge, 38
  corona on video signal, 39, 43
  crosstalk, stereo signal, 167
  distortion factor residual, 190–193
  distortion on sinewave, 125–126
  f.m. detector 'S' characteristics, 123–124
  f.m. tuner response characteristic, 130
  field interval, 19, 36
  field multivibrator waveform, 73
  field period, 20
  field retrace pulse, 70
  field signal on h.t. line, 74
  field sync, 29, 53, 56
  field sync pulse formation, 54, 56

## SERVICING WITH THE OSCILLOSCOPE

Displays *continued*
 field timebase waveforms, 68, 72, 73
  in BRC 9000-series colour receiver, 99
 flyback lines, TV, 64
 flywheel sync, 63-64
 focus electrode, picture tube, signal on, 114
 hum, slight, 19, 65
 hum on TV picture, 45, 47
 hum on video signal, 44
 line blocking oscillator waveforms, 86, 87, 89
 line drive waveforms, 77
 line/field crosstalk, TV picture, 71
 line multivibrator waveforms, 89
 line output transformer ringing waveforms, 93-96
 line output valve screen-grid waveforms, 81-82
 line rings, TV picture, 79
 line sync, 30, 36, 37, 53, 54, 64
 line sync pulses, 20
 line timebase, incorrect speed, TV picture, 84, 86
 line timebase interference, TV picture, 39
 line timebase pulses, 24, 29
 loudspeaker in-room sweep, 201
 magnetic pickup frequency response/ separation, 203
 noise on field sync, 54
 noise on TV screen, 6
 noise signal, 5
 parasitic oscillation, 175-176
 phase oscillograms, 178, 180
 poor h.f. response, 36-37
 rectifier waveforms, 112
 response traces
  a.f., 197-204
  r.f., 119, 122-123, 127
 ripple signal across reservoir capacitor, 112
 square waves, 10
  10 kHz, 105
  audio, 181-188
  video amplifier testing, 105-111
 stereo decoder, 166-167
 stereo f.m. frequency response/ separation sweep, 203
 sweep frequency, audio, 197-204
 sweep linearity check waveforms, 18
 sync separation defect, 57

Displays *continued*
 test card, TV, 37
 test pulses during field sync period, 41
 test signal, TV, 34, 41
 trace distortion check, 18
 vertical non-linearity, TV, 69
 video signal, 40, 51
  colour TV, 150-157
Distortion tests, TV, 102, 189
Dynamic convergence, 145

Electron guns, colour tubes, 146

F.M. tuner alignment, 129
F.M. tuner/receiver response characteristics, 130-131
Field buzz, 75
Field sync, 28-29
Field sync period, 40
Field timebase, 67
 BRC 9000-series colour receiver, 99
Flywheel-controlled line sync, 61-62
Frequency multiplex stereo decoding, 163-164

Gain, audio, 170
Graticule, 2

High frequency response
 good, 37
 poor, 35-36
Hum on TV picture, 45, 47
Hum on video signal, 44
Hum tests
 audio, 186, 188-189
 TV, 102

Inputs, Y, 26
Instruments for sweep alignment, 128, 196
 a.f., 195
 r.f., 115
Integrator, 59-60
Interlacing, TV, 55
Isolation, capacitor, 49

# INDEX

Line blocking oscillator, 83
Line drift, 76
Line drive, 80
Line generators, 83
  faults, 85
Line output stage, 78
Line output transformer, tests, 91–96
Line pulses, 28
Line timebase, 75
  self oscillating, 88, 90
Logarithmic amplifier, 202
Low-frequency response, Y amplifier, 9
Luminance signal, colour TV, 28

Mains hum, 44
Marker distortion (sweep tests), 120
Marker pip (r.f. sweep alignment), 118
Modulation, quadrature, colour TV, 136

Non-linearity, field timebase, 68–69

Oscillograms, 17
  interpretation, 52
  *see also* Displays
Oscilloscope testing in colour receivers, 138

PAL decoder, 142
  waveforms, 141
PAL system of colour TV, 136
Parasitic oscillation, 175–176
Peak-to-peak value of waveform, 27
Peak value of waveform, 27
Phase measurement (oscilloscope method), 176–180
Photography of displays, 17, 22
Pin-cushion distortion and correction, TV, 148–149
Power, average continuous wave, 32
Probe, active, 17
Probe, compensated voltage-divider, 16, 19
Purity, colour TV, 143

Response curve tracing (r.f. and a.f. oscilloscope method), 116, 194
Ringing on square waves, 185

Ripple
  50 Hz, 188
  100 Hz, 189
Rise-time, 8, 17
R.M.S. value of waveform, 27

Shadowmask picture tube, TV, 144
Signal, nature of, 26
Signal generator, TV, use of, 104
Sine-wave signal, 27
Single shot facility, oscilloscope, 12
Slewing rate, how to measure, 193–194
Square wave, harmonic components of, 7
Square-wave testing
  in audio amplifiers, 181
  in video amplifiers, 104–111
Static convergence, colour TV, 145
Stereo decoder, 162
Stereo principles, 158
Stereo radio, 158
  multiplex signal spectrum, 161
  'stereo birdies' interference, 167
  transmitter waveforms, 159–160
Sweep alignment techniques, r.f. and a.f., 115–131, 194
Sweep response, audio, 194
Sweep techniques, 118, 121, 194
Sync facilities on oscilloscope, 9
Sync, line, flywheel-controlled, 61
Sync networks, TV, 58
Sync separator
  Decca colour TV circuit, 50
  valve circuit, 50
Synchronising waveforms, TV, 48–65

Tape recorders, testing in, 171
Test instruments, 21, 196
Test points
  TV sync, 48
  TV video, 46
Testing audio equipment, 169–204
Time-constant, 58
Time multiplex decoding, stereo radio, 163–164
Timebase
  colour TV, 99, 143
  field, BRC 9000-series colour receiver, 99
  oscilloscope, 3
  transistor circuits, testing in, 97

207